AF868434

Zhenjun Ming · Anand Balu Nellippallil ·
Ru Wang · Janet K. Allen · Guoxin Wang ·
Yan Yan · Farrokh Mistree

Architecting A Knowledge-Based Platform for Design Engineering 4.0

Springer

Zhenjun Ming
Institute for Industrial and Intelligent Systems Engineering
School of Mechanical Engineering
Beijing Institute of Technology
Beijing, China

Ru Wang
Institute for Industrial and Intelligent Systems Engineering
School of Mechanical Engineering
Beijing Institute of Technology
Beijing, China

Guoxin Wang
Institute for Industrial and Intelligent Systems Engineering
School of Mechanical Engineering
Beijing Institute of Technology
Beijing, China

Farrokh Mistree
The Systems Realization Laboratory
The School of Aerospace and Mechanical Engineering
University of Oklahoma
Norman, OK, USA

Anand Balu Nellippallil
Department of Mechanical and Civil Engineering
Florida Institute of Technology
Melbourne, FL, USA

Janet K. Allen
The Systems Realization Laboratory
The School of Industrial and Systems Engineering
The University of Oklahoma
Norman, OK, USA

Yan Yan
Institute for Industrial and Intelligent Systems Engineering
School of Mechanical Engineering
Beijing Institute of Technology
Beijing, China

ISBN 978-3-030-90520-0 ISBN 978-3-030-90521-7 (eBook)
https://doi.org/10.1007/978-3-030-90521-7

© The Editor(s) (if applicable) and The Author(s), under exclusive license to Springer Nature Switzerland AG 2022
This work is subject to copyright. All rights are solely and exclusively licensed by the Publisher, whether the whole or part of the material is concerned, specifically the rights of translation, reprinting, reuse of illustrations, recitation, broadcasting, reproduction on microfilms or in any other physical way, and transmission or information storage and retrieval, electronic adaptation, computer software, or by similar or dissimilar methodology now known or hereafter developed.
The use of general descriptive names, registered names, trademarks, service marks, etc. in this publication does not imply, even in the absence of a specific statement, that such names are exempt from the relevant protective laws and regulations and therefore free for general use.
The publisher, the authors and the editors are safe to assume that the advice and information in this book are believed to be true and accurate at the date of publication. Neither the publisher nor the authors or the editors give a warranty, expressed or implied, with respect to the material contained herein or for any errors or omissions that may have been made. The publisher remains neutral with regard to jurisdictional claims in published maps and institutional affiliations.

This Springer imprint is published by the registered company Springer Nature Switzerland AG
The registered company address is: Gewerbestrasse 11, 6330 Cham, Switzerland

Foreword

Architecting Knowledge-Based Platform for Design Engineering 4.0

The advent of the internet and advances in cloud computing have spawned a new paradigm for engineering—a network-centric environment of cooperating engineering applications (Industry 4.0 or Industrie 4.0). This new paradigm involves symbiotic networks of people (social networks), services, and smart devices with sensing capabilities; these are also referred to as the Internet of People, Internet of Services, and Internet of Things, respectively. To realize this vision of globally cooperative engineering, we need to address several computational challenges: engineering design, modeling, interoperability, knowledge/information/data analytics (involving artificial intelligence techniques [AI]), privacy, security, network behaviors, visualization, and novel architectures and storage mechanisms. The authors of this monograph focus on several of these topics, in particular, engineering design, modeling with ontologies, decision-making using AI techniques that are instantiated in a cloud-based architecture.

The authors briefly discuss the evolution of engineering design paradigms in line with the evolution of various industrial revolutions—from Industry 1.0 to Industry 4.0. In Industry 1.0, the focus of design was not on customers but more on mechanization. However, in Industry 4.0, it is customer and network-centric. The authors augment the Industry 4.0 construct to include a new paradigm for designing, namely, Design Engineering 4.0, that is appropriate use in a network-centric environment of cooperating engineering applications.

The senior authors of the monograph—Farrokh Mistree and Janet K. Allen—are known for their contributions to engineering design. They have pioneered a technique called Decision Support Problem Technique (DSPT) and its extensions. A satisficing strategy instead of an optimization strategy is implemented in the DSPT. In this monograph, the authors extend their previous work to support knowledge-intensive collaborative design, taking into consideration various uncertainties that arise in design. They describe the architecture and functionalities of a platform to support knowledge-intensive design—Platform for Decision Support in the Design of Engineered Systems (PDSIDES). PDSIDES supports many of the functionalities needed for Design Engineering 4.0—knowledge management, decision-making, workflow to support collaborations, uncertainty management, and adequate user support.

Foundational to architecting the knowledge-based platform is the ontology for selection, compromise, and hierarchical coupled decisions and the Decision Support Problem template (addressed in Chaps. 2 and 3). The authors demonstrate the efficacy of the foundations for different users to formulate and capture decision-related knowledge, thereby leading to the development of the PDSIDES prototype (addressed in Chap. 4). Further, the authors extend the foundational ontology to support users in carrying out the meta-design of decision workflows. Ontology to represent the knowledge in decision workflows based on the Phase-Event-Information-X (PEI-X) diagram is developed by the authors as part of this functionality (addressed in Chap. 5). To facilitate uncertainty management in decision workflows and capture the associated knowledge, the authors present the ontology for a systematic robust design space exploration procedure (addressed in Chap. 6). Users thereby are able to formulate decision-centric design workflows under uncertainty and explore design alternatives that are relatively insensitive to uncertainty—defined as "satisficing robust solutions."

Cloud computing, where services are delivered over the Internet, has the potential to transform the practice of engineering. With tools and services residing in the cloud environment, engineering and manufacturing companies can now access shared computing resources and advanced application services everywhere and anytime on an as-needed basis. This results in considerable cost savings in expenditures on Information Technology (IT) and enhances the ability to deliver high-quality engineered products. In the final chapter (Chap. 7), the authors describe a cloud-based extension to PDSIDES—called CB-PDSIDES. This extension includes intelligent services composition, security, and other tools and techniques for realizing the Internet of X, discussed earlier. The authors also discuss the various challenges involved in realizing such an environment.

I am pleased to recommend this stimulating and insightful monograph for your personal library, organization, university, or company (whether Fortune 500 or a start-up). I would also encourage the ideas presented here to become part of a broader conversation regarding design engineering at the interface of various engineering disciplines. Their approach to architecting a knowledge-based platform to support design engineering 4.0 is systematic and well-conceived, helping a reader organize the relevant challenges and opportunities for transformational change. I trust its value will endure.

Ram D. Sriram, Ph.D.
Chief Software and Systems Division, Information Technology Laboratory
National Institute of Standards and Technology
Maryland, USA
https://www.nist.gov/people/ram-d-sriram

Preface

Design Engineering for Industry 4.0 (DE4.0) represents the 'human-cyber-physical view of the systems realization ecosystem' that is necessary to accommodate the drivers of Industry 4.0 (IoX) and provide an open ecosystem for the realization of complex systems. Seamless integration of digital threads and digital twins throughout the product design, development, and fulfillment lifecycle; ability to accommodate diverse and rapidly changing technologies; mechanisms to facilitate the creation of new opportunities for the design of products, processes, services, and systems are some of the desired characteristics of DE4.0.[1]

In keeping with the Design Engineering 4.0 construct, we describe architecting a computer platform to support human designers make decisions associated with the realization of complex engineered systems. The platform is designed to facilitate end-to-end digital integration, customization and personalization, agile collaboration networks, open innovation, co-creation and crowdsourcing, product servitization, and anything-as-a-service.

Recognizing that simulation models are abstractions of reality, we opt for a satisficing strategy instead of an optimization strategy. We include fundamentals and describe tools for architecting a knowledge-based platform for decision support. Challenges associated with developing a computational platform for decision support for the realization of complex engineered systems in the context of Design Engineering 4.0 are identified. Constructs for formulating design decisions (e.g., selection, compromise, and coupled decisions), knowledge modeling schemes (e.g., ontologies and modular templates), diagrams for designing decision workflows (e.g., the PEI-X diagram), and some analytical methods for robust design under uncertainty are presented. We describe integrating the knowledge-based platform with the cloud to

[1] Jiao, R., Commuri, S. Panchal, J., Milisavljevic-Syed, J, Allen, J.K., Mistree, F. and Schaefer, D., "Design Engineering in the Age of Industry 4.0," *ASME Journal of Mechanical Design*, 143(7), 070801, 25 pages.

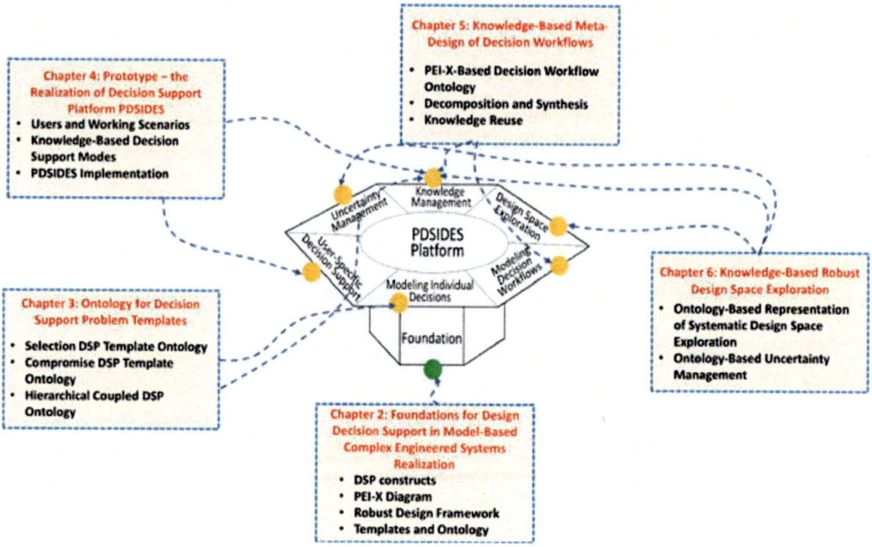

Fig. 1 Challenges addressed in the monograph

architect a cloud-based platform for decision support promoting co-design and cloud-based design communication essential for mass collaboration and open innovation for Design Engineering 4.0.

The key features of the knowledge-based platform PDSIDES for Design Engineering 4.0 include

- Capability to integrate models and simulation tools spanning different processes and length scales (typically defined as vertical and horizontal integration from the context of integrated design of materials, products, and manufacturing processes),
- Capability to define computational workflows involving decision-making, spanning multiple activities and users,
- Capability to define modular, reusable sub-workflows for specific processes,
- Capability to connect to external databases on materials, products, and processes,
- Capability to provide knowledge-guided assistance to different types of users in design-related decision-making,
- Capability to carry out collaborative and multidisciplinary design,
- Capability to manage complexity (reduced cost of computation via surrogate models/metamodels),
- Capability to explore and visualize the design and solution space,
- Capability to carry out dynamic and cost-efficient reconfiguration and integration of design decision templates to explore different robust design strategies (meta-design to deliver robust products).

Major challenges addressed in this monograph are illustrated in Fig. 1. They include

- Foundational concepts and techniques for architecting a knowledge-based decision support platform,
- Ontologies for individual decisions, including selection, compromise, and coupled decisions,
- The functionalities (components) of the PDSIDES platform,
- Ontology for meta-design of designing decision workflows,
- Ontology for robust design space exploration.

This monograph is comprised of seven chapters; see Fig. 2. In Chap. 1, we provide an overview of the monograph, and in Chap. 7 a summary of what is presented and more importantly identify in the context of Internet of People, Internet of Services, Internet of Things, and Internet of Commerce areas for future research for furthering Design Engineering 4.0. These include

- Design for User Experience,
- Design of Human-Cyber-Physical Systems,
- Design with Smart Sensing and Artificial Intelligence Technologies,
- Design as Strategic Engineering.

While we recommend reading the chapters sequentially, each chapter is self-contained, and chapters can be read independently or in any preferred sequence.

In Chap. 2, we introduce the foundations for the decision support platform architecture, namely, (1) primary constructs in decision-based design, (2) frameworks for robust decision-making, (3) the PEI-X diagram for designing decision workflows, and (4) knowledge-based techniques for decision support. In Chap. 3, we address modeling and capturing knowledge related to selection, compromise, and coupled hierarchical decisions in design using template-based ontological methods. In Chap. 4, we present the knowledge-based platform PDSIDES, its users and working scenarios, and the corresponding knowledge-based decision support modes. The efficacy of the platform is tested using a multiscale, multistage materials design problem—vertical and horizontal integration and integrated design of hot rod rolling process chain (manufacturing processes), steel (material), and rolled rod (product). In Chap. 5, we extend the functionalities of PDSIDES presented in Chap. 4 to facilitate capturing, representing, and documenting the knowledge related to hierarchical decision workflows in designing the process of designing complex engineered systems, that is, the meta-design of complex systems. The meta-design of a heat exchanger in a thermal system is used as an example to test the functionality. In Chap. 6, the functionalities of PDSIDES are extended to include uncertainty management and formulating robust design decision workflows. The ontologies for robust design and a template-based ontological method that facilitate the systematic formulation of decision workflows under uncertainty are presented. In Chap. 7, we summarize our contributions in this monograph, propose a conceptual framework for a Cloud-Based Platform for Decision Support in the Design of engineered Systems (CB-PDSIDES) and identify several new research opportunities in Design Engineering 4.0, and provide some closing comments.

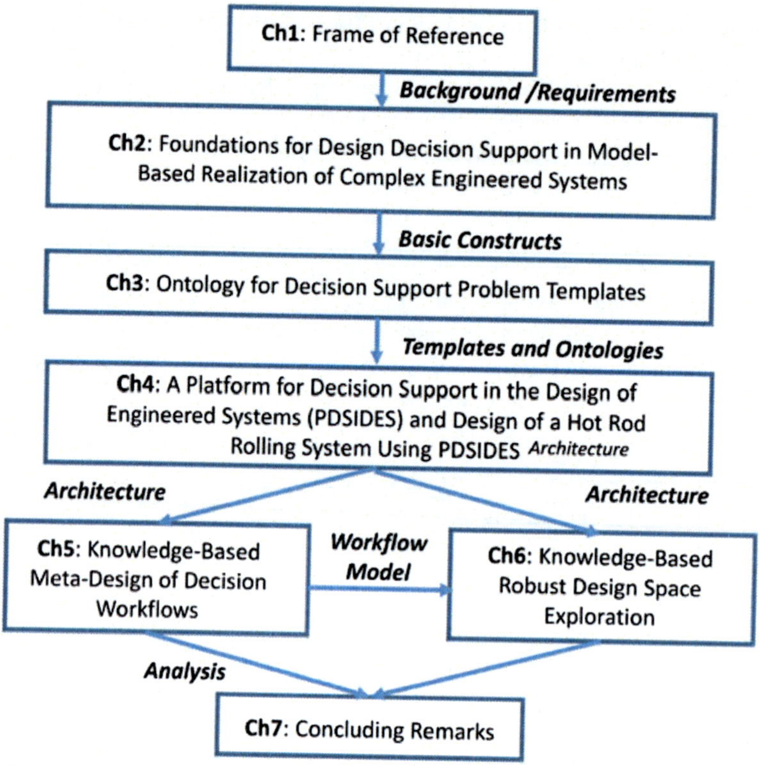

Fig. 2 Organization of the monograph

In closing, we observe that the work embodied in this monograph is one of the outcomes of the Systems Realization Partnership conceived by Professor Yan Yan and ably instantiated and nurtured by Professor Guoxin Wang. The material in this monograph is anchored in the doctoral dissertations of three former doctoral students and their collaboration among each other and their faculty mentors:

- Zhenjun Ming and Ru Wang, whom we were privileged to mentor with our colleagues Professors Guoxin Wang and Yan Yan from the Beijing Institute of Technology, Beijing, China.
- Anand Balu Nellippallil, whom we were privileged to mentor together with our colleagues Dr. B. P. Gautham (TCS Research, Pune, India) and Professor Amarendra Kumar Singh (IIT Kanpur, India).

We recognize Maryam Sabeghi and Chung-Hyun Goh who collaborated with Ming and Wang and Vignesh Rangaraj, who collaborated with Nellippallil during his doctoral research and added value to the material presented herein. We received financial support from several sources to pursue the concepts outlined in this monograph. Ming and Wang were supported by scholarships from China Scholarship Council and Beijing Institute of Technology Postgraduate Internationalization Project Foundation. Funds from the LA Comp Chair account and the John and Mary Moore Chair account at the University of Oklahoma, and the Tata Consultancy Services, India, were used to support Nellippallil through his doctoral studies in the School of Aerospace and Mechanical Engineering at the University of Oklahoma, Norman, USA.

Norman, USA
September 2021

Janet K. Allen
Farrokh Mistree

Contents

Chapter 1
Requirements and Architecture of the Decision Support Platform for Design Engineering 4.0

In the era of Industry 4.0, with the increase of machine intelligence and the internet of things/services in production systems, one of the ramifications is that it will foster a highly customized and dynamically changing market where more and more customers demand personalized or individualized products. This imposes a challenge to designers who are the decision-makers in the design of these products as well as the systems that support the production of the products. Designers must be able to quickly respond to market variations and make quality decisions. This necessitates a platform for supporting design decisions. To assist in designing and developing such a platform, in this chapter we introduce the paradigm of Industry 4.0 and analyze the features of design decisions and decision workflows in the context of Industry 4.0. Then, we identify the requirements of a platform to support these decisions and decision workflows, and we propose an architecture that meets the requirements of the platform. Finally, we outline the organization of the monograph.

1.1 Background: Design Decision Support in the Context of Industry 4.0

The fourth industrial revolution, also known as Industry 4.0, is significantly influencing manufacturing industries across the globe [1]. Industry 4.0 represents a comprehensive transformation of the whole sphere of industrial production through the merging of digital technology and the internet with conventional industry [2]. It provides a framework to address the challenges arising in the integration of cyber-systems and physical resources and covers all aspects of manufacturing systems [3], including: robust and flexible automation; data collection, analysis, learning and decision-making; distributed production systems; industrial IoT; and supply chain integration [4]. Factories conforming to Industry 4.0 will integrate services across the entire manufacturing and operations processes and will be able to adapt to disruptions in real-time, thereby improving the quality of products and services [5].

© The Author(s), under exclusive license to Springer Nature Switzerland AG 2022
Z. Ming et al., *Architecting A Knowledge-Based Platform for Design Engineering 4.0*,
https://doi.org/10.1007/978-3-030-90521-7_1

The digital transformation of manufacturing industries brought about by Industry 4.0 has created a framework through which substantial improvements in productivity, quality, and customer satisfaction can be achieved. This digital transformation, according to Jiao et al. [4], not only affects the way products are manufactured, but also creates new opportunities for the design of products, processes, services, and systems. Engineering design is essentially a decision-making process and the principal role of designers is to make decisions [6]. Providing decision support is critical for augmenting human designers' decision-making ability and thus generating quality designs. In this section, we review the literature related to design, decisions, and decision support, which provides the background for the need of a platform for decision support in the design of engineered systems in the context of Industry 4.0.

1.1.1 Design Engineering 4.0 and the Industrial Brain

Engineering Design, by the definition of Accreditation Board for Engineering and Technology (ABET) [7], is the process of devising a system, component, or process to meet desired needs. It is a decision-making process (often iterative), in which basic science, mathematics and engineering sciences are applied to convert resources optimally to meet a stated objective. Among the fundamental elements of the design process are the establishment of objectives and criteria, synthesis, analysis, construction, testing, and evaluation.

The evolution of engineering design, as shown in Fig. 1.1, is typically divided into four stages (i.e., Design Engineering 1.0 to 4.0) according to its interplay with the industrial processes of manufacturing from Industry 1.0 to 4.0 [4]. In the age of Industry 1.0, industries tried to use steam power to offload labor-intensive manufacturing processes, and the focus of design and manufacturing was on mimicking

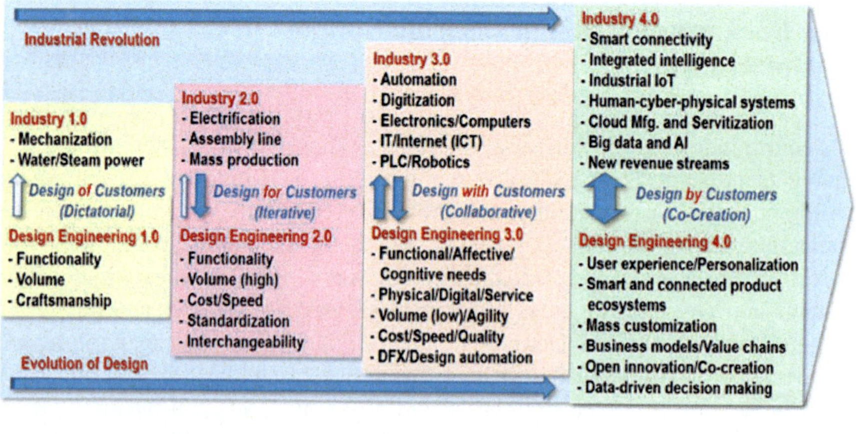

Fig. 1.1 Design Engineering 4.0 in line with Industry 4.0 [4]

the process adopted by the human worker. The key feature of Design Engineering 1.0 is dictatorial and can be summarized as "design of the customers". In the age of Industry 2.0, electrification and assembly lines gave birth to the mass production of products so that products could be produced in large volumes and at higher rates. Therefore, the focus of Design Engineering 2.0 shifted to product assembly which led to the push for standardization and interchangeability of parts/processes to achieve economies of scale. And redesign is often needed to address issues in manufacturing. The focus of Design Engineering 2.0 is "design for customers", and it is an iterative process. Industry 3.0 is characterized by increasing automation and the use of CNC machines. Manufacturing processes are networked, and data can be shared across processes to ensure high-quality products with tight tolerances. Because of digitalization, CAD designs can be directly used in the manufacturing process using CNC Machines, and design and manufacturing became more tightly coupled than ever before. Therefore, the key feature of Design Engineering 3.0 is being collaborative, and can be summarized as "design with customers". In the age of Industry 4.0, manufacturing systems are expected to able to add product personalization to a mass-produced product, which means that designers must partition base functionality of the product from customizable features. Engineering Design is evolving to a new paradigm for design by customers through co-creation of product value chain fulfillment in a human-cyber-physical environment [8]. Jiao et al. [4] define this new paradigm as Design Engineering 4.0, which represents the "human-cyber-physical view of the systems realization ecosystem" that is necessary to accommodate the drivers of Industry 4.0 (IoX) and provide an open ecosystem for the realization of complex systems. The embodiment of human-cyber-physical view of the systems realization ecosystem represented by Design Engineering 4.0 is shown in Fig. 1.2. The key idea is anchored in: (1) the Internet of People, where designers take into account user preferences and users can interact with the products and among themselves; (2) the Internet of Services, where products and services are customized to user requirements while producing products of "zero lot size" and "mass production costs"; (3) the Internet of Commerce, where business are run by monetizing services through the internet and design is performed as strategic engineering; and (4) the Internet of Things, where systems are designed with smart sensing and AI technologies.

The concept of "human-cyber-physical system" is also found in Zhou et al.'s [9] work where they attempt to define the new generation of intelligent manufacturing (NGIM). In their framework, they define NGIM as a composite intelligent system that is comprised of relevant humans, AI-capable cyber-systems, and physical systems, with the aim of achieving specific manufacturing goals at an optimal level. Physical systems act as the "executing body" which execute the energy and material flow of manufacturing activities. AI-capable cyber-systems act as the core of the information flow of manufacturing activities, and help humans to complete the necessary perception, cognition, analysis, decision-making, and control of the physical systems for optimal operation. The human is the "master" who create both the physical systems and the cyber-systems, and remains in the central position to make decisions and enact control. The advance of NGIM is anchored in the idea of using AI-capable

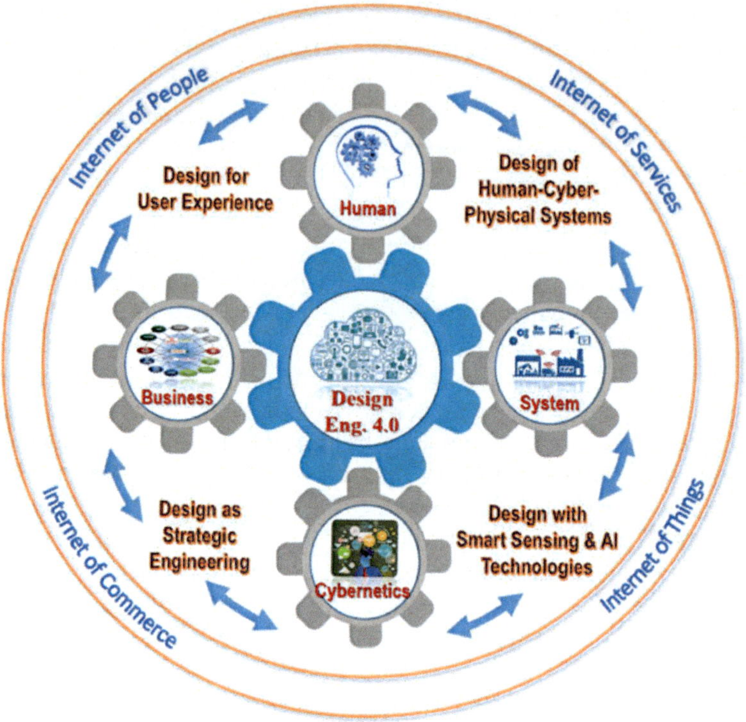

Fig. 1.2 Design Engineering 4.0: human-cyber-physical view of the systems realization ecosystem [4]

cyber-systems to offload a substantial amount of mental work from humans, given that most of the manual labor work has been replaced by physical systems. This idea is also emphasized by Kang et al. [10] who propose the so-called "Industrial Brain" as the reference model for realizing intelligent manufacturing. The "Industrial Brain" is comprised of big data, computing power, and algorithms, and its main functionality is to replace the mental work of humans in decision-making. Since engineering design is essentially a decision-making process and most of the designing work is in the mental domain, the "Industrial Brain" seems a promising intelligent "HUB" for supporting Design Engineering 4.0. However, one of the disadvantages of the "Industrial Brain" is related to its ultimate goal of replacing the role of decision-making of humans, which is not possible since human judgement is critical in design. To address this, we revise the "Industrial Brain" model and tie it to the "human-cyber-physical system" framework, as shown in Fig. 1.3.

Using the data collected from the physical system (consist of materials, machines, and products, etc.) through advanced sensors, the algorithms and computing power provided by the cyber-system (consisting of digital twins, digital threads, cloud computing, etc.), and the judgement provided by the human beings (e.g., designers, manufacturers, suppliers, and customers, etc., who are the key stakeholders in the

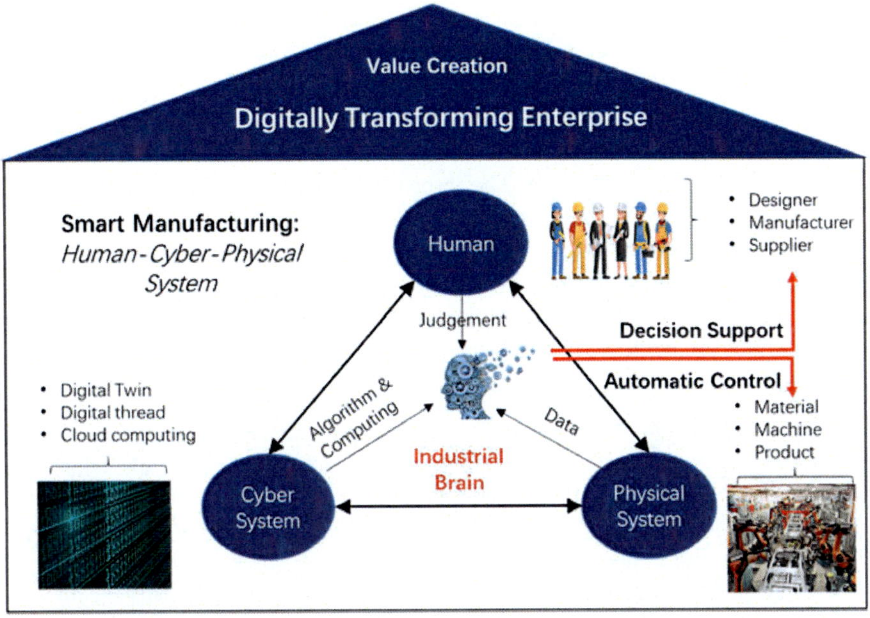

Fig. 1.3 "Industrial brain" of "human-cyber-physical systems"

manufacturing business), the "Industrial Brain" is expected to deliver real-time control to the physical system and real-time decision support to human beings. The idea of decision support is very important in the context of Design Engineering 4.0, which is discussed next in Sect. 1.1.2.

1.1.2 *Decisions and Decision Support in the Context of Design Engineering 4.0*

As mentioned earlier, one of the ramifications of Industry 4.0 due to the digitization of every part of a manufacturing enterprise and connecting them using the internet of things and the internet of services is that it will enable flexible production of small batch products and foster a highly customized and dynamically changing market in which more and more customers demand personalized or individualized products. To respond to this trend, as what is stated in Jiao et al.'s definition of Design Engineering 4.0 [4], we need a "human-cyber-physical view of the systems realization ecosystem" that can accommodate customers' individualized demands and develop proper fulfillment capabilities. The human dimension of Design Engineering 4.0 highlights the importance of user experience not only for customers as end-users at the frontend, but also for multiple stakeholders especially designers at the backend of product realization [4]. Mistree et al. [6] point out that the principal (but not only) role of a designer is to make decisions. It is critical to improve the experience of a design decision-maker in the age of Design Engineering 4.0 so as to speed up the decision-making process and generate quality designs.

Hazelrigg [11, 12] defines a decision as an irrevocable allocation of resources, which is made at an instant in time and must be made based on the information available at the time it is made. The information is often from many sources (and disciplines) and the decisions that are made based on such information may have wide-ranging repercussions. Decisions help bridge the gap between an idea and reality. In Decision-Based Design, decisions serve as markers to identify the progression of a design from initiation to implementation to termination [13]. Some principal observations and beliefs from a Decision-Based Design perspectives are as follows [6]:

- The principal role of an engineer or designer is to make decisions.
- Design involves a series of decisions some of which may be made sequentially and others that must be made concurrently.
- Design involves hierarchical decision-making and the interaction between these decisions must be taken into account.
- Design productivity can be increased through the use of analysis, visualization and synthesis in complementary roles, and by augmenting the recognized capability of computers in processing numerical information to include the processing of symbols (for example, graphs, pictures, drawing, words) and reasoning (for example, artificial intelligence).
- Life-cycle considerations that affect design can be modeled in upstream design decisions.

The characteristics of decisions are governed by the characteristics associated with the design of real-life engineering systems. These characteristics are summarized by the following descriptive sentences:

- Decisions in design are invariably multileveled and multidimensional in nature.
- Decisions involve information that comes from different sources and disciplines.
- Decisions are governed by multiple measures of merit and performance.
- All the information required to arrive at a decision may not be available.
- Some of the information used in arriving at a decision may be hard, that is, based on scientific principles and some information may be soft, that is, based on the designer's judgment and experience.
- The problem for which a decision is being made is invariably loosely defined and open. Virtually none of the decisions are characterized by a singular, unique solution. These decisions are less than optimal and are called satisficing solutions.
- Design is the process of converting information that characterizes the needs and requirements of a system into knowledge about the system itself. In Decision-Based Design, it is the making of decisions that brings about the transformation of information into knowledge about a system.

In addition to the characteristics summarized above, decisions in engineering design are also characterized by ambiguity, uncertainty, risk, and tradeoffs [14].

Due to these characteristics, making design decisions in a rigorous way is nontrivial even though the fundamental principles of decision analysis are generally applicable to decision-making in engineering design [15]. Taking into consideration the new features of Industry 4.0 such as dynamic market and personalized customer demands, designers as decision-makers are facing new challenges in the design of products themselves as well as the processes for realizing them. These challenges include dealing with: (1) frequent and dynamic variations in both products and the associated processes, (2) the intensive, multidisciplinary knowledge that is needed in design, (3) the uncertainties that are anchored in the parameters and models that are used for decision-making, (4) the huge amount of (user generated) data/information that needs to be processed to generate useful results for decision support. Confronting these new challenges, as shown in Fig. 1.4, we believe good decisions in engineering design should be: (i) *rapid decision,* which can quickly generate some results to respond to changes; (ii) *informed decisions*, which are made based on gathering sufficient information/data; (iii) *satisficing decisions*, which are good enough answers to questions whose best answers are unknowable (for the differences between *optimizing* and *satisficing* strategy, see Sect. 2.1.2 in Chap. 2); (iv) *robust decisions*, which are relatively insensitive to variations; and (v) *visualized decisions*, which are made based on that information are shown in a visualized, meaningful, and intuitive way. To realize this, there is an emerging need for decision support tools, platforms, or an ecosystem (e.g., the "Industrial Brain") in design.

Recently there has been a trend to support design decisions from a data-driven perspective, where most of the data are user experience data which are collected from smart sensing [16], E-commerce platforms [17], social media [18], etc. The basic idea behind data-driven decision-making is to base design decisions on facts instead of assumptions [4]. For example, Young et al. [19] use the data acquired from physiological experiments to support decisions in the design of robotic control systems. Green et al. [20] leverage the product usage data collected by customer interviews to gain insights for product design. Zhang et al. [21] propose a method for projecting actual product operating data to the design stage so as to support decisions

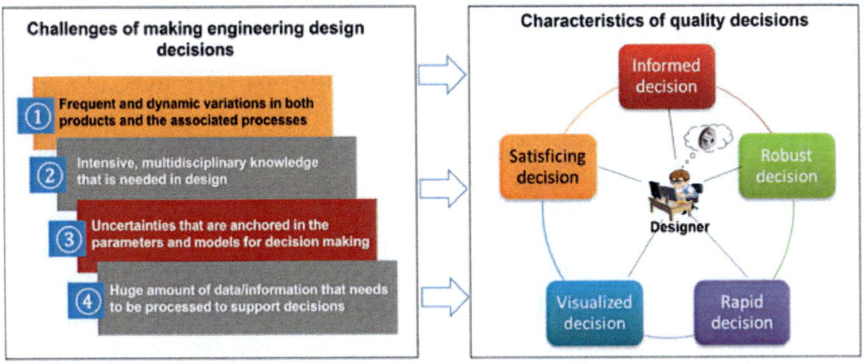

Fig. 1.4 Characteristics of quality decisions confronting the challenges

in early design stages. Zhou et al. [22] propose a method that can extract latent customer needs from online product reviews through use case analogical reasoning, which is helpful to support decisions in design. Zheng et al. [23] propose a framework for utilizing the data generated by users to support design innovation in a cloud-based cyber-physical environment.

The essence of the prevailing data-driven decision support methods is to use the user-generated content or experiment and simulation data to create surrogate (or approximate) models that can be used to support decisions [4]. The challenge is that the design process itself and the decisions involved cannot be driven by this data per se. What really supports designers to make informed decisions is the design knowledge and informatics acquired from data [21]. Therefore, data-driven design can be essentially viewed as a knowledge-based design decision-making process [24]. This is consistent with the notion that design decision support should emphasize knowledge management and value extraction to achieve an integration from data to information and then to knowledge in light of the DIKW pyramid model [25].

In summary, knowledge plays a critical role in supporting design decisions and there is a need of platform that can make full use of knowledge to support human designers in decision-making. To address this need, in this monograph we propose to architect a knowledge-based Platform for Decision Support in the Design of Engineered Systems (PDSIDES). Specific requirements of such a platform are discussed in Sect. 1.2.

1.2 Requirements for a Design Decision Support Platform

1.2.1 Knowledge Management and Reuse

Decision-making in design is a knowledge-intensive process which requires different types and sources of knowledge as input. Most of the tasks that designers perform in design require applying scientific and engineering knowledge to find (generate, evaluate and select) the solutions to design problems, and then improve or optimize those solutions within the framework composed of requirements and constraints set by physical, environmental and human-related considerations [26]. Building a knowledge base that manages the design knowledge is critical for supporting decisions. There are two types of knowledge: knowledge about the design process (procedural knowledge) and knowledge about the product (declarative knowledge) being designed. According to Rich [27], procedural knowledge is defined as a set of well-defined procedures that represent information about doing things, and declarative knowledge is defined as a set of facts represented (usually) according to the protocol defined by procedural knowledge. In the context of design decision-making, procedural knowledge dictates how information is to be used and depends on the design method and not its application. The information about structuring, defining, and processing the design problem is and does not change. This procedural knowledge is domain-independent. It is the knowledge about using knowledge and must be modeled in the knowledge base. Declarative knowledge is dependent on the product being designed and independent of the method used to design it. This type of knowledge represents the physics of the problem, serves as the input of procedural knowledge, and must also be modeled in the knowledge base for supporting design decisions.

Modeling design decision knowledge in a reusable manner is another important factor for supporting decisions in addition to knowledge management. Baxter et al. [28] pointed out that knowledge reuse can bring many benefits to enterprises for improving their engineering designs, because the next generation product is likely to have a significant overlap with the previous version, and knowledge reuse allows more time for innovation especially when it is difficult to achieve a competitive advantage in mature domains. In the context of decision-making, overlaps are often seen in many decisions where variant or adaptive design is the case and knowledge reuse can save the designer a significant amount of time in making the decision. Therefore, we need a functionality of knowledge reuse in the platform to improve efficiency in making decisions.

1.2.2 Formulation of Decisions and Decision Workflows

According to the definition in Wikipedia [29], formulation is to put together of components in appropriate relationships or structures, according to a formula. In that sense a formulation is created according to the standard for the product. In the context of engineering design, formulation of decisions means framing or structuring the components of a design problem in a way that facilitates a designer making a rational decision about it. Mullen and Roth [30] describe decision-making as a process that includes recognizing problems and analyzing values; generating alternative choices (gathering information about choices); evaluating choices (identifying best choices [optimizing] or satisfying some external criteria [satisficing] by analyzing cost and benefits of outcomes); and binding the will (committing to choice) and ignoring sunk cost (effort already expended). In the light of this description, the formulation of a design decision is to cast the design problem into those sub-processes, and the formulation should have some mathematics or quantification metrics to be involved so as to ensure mathematical rigor. To support the formulation of decisions, a computer-based platform needs to have linguistic and mathematical constructs (serving as formulas) so that designers can directly use them for structuring the problem as decisions.

Even though Mullen and Roth [30] have defined a generic decision-making processes, the formulation of design decisions can be more complex than that. The first complexity is anchored in that the alternative choices in design can be discrete, continuous, or a mix, and the number of choices can be very large (to infinity). For example, in the design of a cantilever beam, alternatives include a set of materials (discrete), a set of cross-section shapes (discrete), and the sizes (continuous). The overall design candidates are a combination of all the three, which make formulating the choices a complex task. The second complexity is anchored in the problem that the evaluation of choices needs to consider both objectives and constraints which can be linear or nonlinear functions of the choice attributes. For example, when evaluating an alternative design of a cantilever beam, a designer needs to not only consider minimizing the mass as a goal but also consider the constraints introduced by yield strength and bending strength, etc. These are nonlinear functions of the beam properties. The third complexity is anchored in the formulating of uncertainty and risk of choices (which is discussed in Sect. 1.2.4) before committing to them. For example, in a cantilever beam design problem, what are the uncertainties associated with the choices, and what are the associated risks if a designer chooses them for implementation? These concerns must be well addressed in the decision support platform.

In addition to the formulation of individual or single decisions (e.g., the design of a cantilever beam), the platform also needs to support the formulation of decision workflows which represent a network of individual decisions. Engineering systems especially complex systems are, by definition, made-up of inter-related subsystems [31]. For example, a gearbox is comprised of multiple gears and shafts which are assembled to a product which performs a function of transferring motion and torque. The design of such complex systems involves a network of decisions to be made. There are usually coupling or dependent relationships among these decisions, which may result in the consequence that a decision influences other decisions and vice versa. For example, the decision about the material and sizing of a shaft will impact the decisions about the gear and vice versa, because they are strongly coupled. The workflows and the associated interactions among decisions are frequently seen in the design of complex systems, and must be addressed in the decision support platform.

1.2.3 Solution Space Exploration

As we mentioned in Sect. 1.2.2, in the design of engineering systems the alternative choices can be infinite (especially when the choices or design variables are continuous), which means the solution space for a decision can be huge and multidimensional. Therefore, there is a need to support designers to explore the solution space and search for those solutions that can satisfy the goals as well as possible.

Kang et al. [32] identify three ingredients of design solution space exploration, namely, representation of the solution space, analysis of the solutions, and the exploration methods. The representation of the solution space is to formally define a space in which every design is valid and satisfies all the constraints. This can be done by identifying the bounds of design variables, the function, and limits of constraints. The solution analysis means analyzing the solutions against the constraints and goals using computational methods such as finite element analysis or surrogate models (e.g., response surfaces that approximate the response of a design given an input).

There must be some computational tools to facilitate designers performing the overwhelming analysis tasks so as to improve efficiency. Exploration methods are used to search for superior solutions and these methods involve search or optimization algorithms. There are many exploration methods in the optimization literature, for example, Tabu Search [33], Genetic Algorithms [34], Rapidly-exploring Random Trees [35], simulated annealing [36], the classic mathematic programming [37], and goal programming [38]. The decision support platform should have exploration techniques as built-in functions to support exploration of the solution space.

In addition to the three basic ingredients, solution space exploration also needs to account for variations (e.g., the variation of the boundary) that may happen to the solution space caused by uncertainties. This is discussed in Sect. 1.2.4.

1.2.4 Uncertainty Management

Uncertainty management in the context of supporting design decisions is to inform designers about the risk of making decisions under uncertainty, facilitate designers managing (not removing) the uncertainty and searching for robust solutions that are insensitive to uncertainty. In model-based or simulation-based design, decisions are usually based on models with simplified assumptions, idealizations, and non-deterministic simulations which inevitably bring about uncertainty. As George Box [39] said: "essentially, all models are wrong but some models are useful," designers have to accept the fact that the models used to support decisions are not perfect. In the real-world practice, it is expensive or even impossible to eliminate all sources of uncertainty, and designers have to manage the uncertainty and identify a way to find solutions which are robust or insensitive to variations. To well manage uncertainty, having a classification scheme is very important.

In the engineering design literature, many classifications and definitions have been proposed. For example, Isukapalli et al. [40] classify uncertainties into three types: (1) "natural uncertainty/variability" which stands for inherent randomness or unpredictability of the physical system; (2) "model uncertainty" which refers to approximations and simplifications in model formulation, and (3) "data uncertainty" denoting incomplete knowledge of model parameters/inputs. Based on Isukapalli's classification, Choi et al. [41] further introduces propagated uncertainty caused by a chain of models in the context of hierarchical system design. Oberkampf et al. [42] carefully distinguishes between variability, uncertainty, and error in computational simulations. In their definition, variability is defined as inherit variation associated with the physical system; uncertainty is defined as a potential deficiency in any phase or activity of the modeling process that is due to lack of knowledge; error is defined as a recognizable deficiency in any phase or activity of the modeling process that is not due to the lack of knowledge. From a system function perspective, Allen et al. [43] classify uncertainty into three categories, namely, uncertainty in noise or environment, uncertainty in design variables or control factors, and uncertainty introduced by modeling methods. Based on this classification scheme, three types of robust design are identified accordingly.

Since uncertainty is inevitable, expensive or impossible to eliminate, we need a functionality in the decision support platform to facilitate designers in effectively managing uncertainty and making robust decisions.

1.2.5 User/Activity Specific Decision Support

User-specific decision support means that the platform should be able to provide decision support to different types of users (decision makers) in various activities. Herbert Simon [44] has classified decision-making activities into three categories: intelligence, design, and choice. Intelligence, or problem finding, includes activities such as comparisons of current status with goals or standards, exception reporting, preliminary computations, and so on. Design encompasses activities related to development of alternatives. Choice covers activities related to evaluating and selecting from alternatives. These activities can find their counterparts in engineering design, such as gathering information and framing the design problem, generating design concepts, evaluating and choosing a design concept for embodiment design, etc. Due to the different features of the activities, the needs of support are also different, which requires the platform to have the flexibility to accommodate the variety of decision activities. In addition to the difference in activities, the styles, skills, and knowledge of decision-makers may also be different [45]. This observation is particularly true in engineering design. For example, designers can be roughly classified into three types based on their level of knowledge and experience, namely, expert designers, senior designers, and novice designers. Due to the difference in knowledge and experience, what they need from a platform to support their decisions varies. An expert designer has considerable knowledge about design and can perform the decision-making process independently, the support that he or she needs from the platform is very different from a novice designer who only has basic knowledge about design and needs to get most of the knowledge from the system. Therefore, a decision platform needs to accommodate the variety in decision-makers.

1.3 Architecture and Functionalities of the Design Decision Support Platform

In order to fulfill the requirements identified in Sect. 1.2, we propose an architecture for the platform PDSIDES as shown in Fig. 1.5. It is an open architecture that is rooted in the foundation of Decision-Based Design and Knowledge-Based Systems, and will provide designers with decision support in the design of engineered systems by

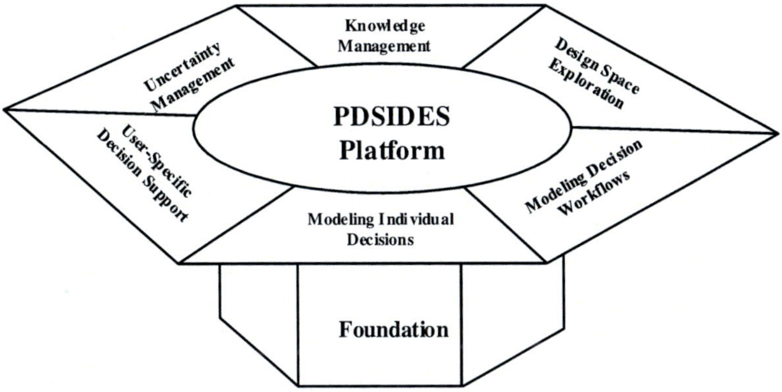

Fig. 1.5 The architecture of the platform PDSIDES

exerting some key functionalities. We summarize the key features and functionalities as follows:

- ***Goal***. Our goal in architecting the platform PDSIDES is to augment (not to replace) the designers' role of decision-makers in the design of complex engineered systems so as to improve the efficiency of the decision-making process and the quality of the decisions that are made. This is different from some design automation platforms which attempt to automate the entire design process including decision-making and output a solution which is the final design. In PDSIDES, a designer is the "master" who makes the decision, the computer is an assistant who provides support that is needed by the designer to make the decision.
- ***User***. Users of PDSIDES are designers who need support to make *informed, rapid, satisficing, visualized*, and *robust* decisions when designing engineered systems.
- ***Foundation***. The foundation of the platform is two-fold: decision and knowledge. From the decision aspect, PDSIDES is based on the Decision Support Problem Techniques (DSPT) [46] that provides the basic constructs for modeling decisions including selection and compromise decisions and combinations thereof (i.e., couple decisions). Derivatives of the DSPT include the Phase-Even-Information-X (PEI-X) diagram [47] for modeling decision workflows, and the robust design framework [43] for managing different types of uncertainties. To model knowledge we use templates and ontology [48] to represent decision-related knowledge and build the knowledge base for providing decision support. Details of the foundation are discussed in Chap. 2.
- ***Functionalities***:
  - ***Knowledge Management***. This functionality ensures that the decision-related knowledge (including both procedural and declarative knowledge) is well managed in PDSIDES. Knowledge is captured when designers use templates for making decisions in design, the captured knowledge is then stored in a knowledge base which is built with ontologies to insure that populated knowledge base is searchable, sharable, and reusable.

- ***Individual Decisions Formulation***. This functionality facilitates designers formulating single decisions using the Decision Support Problem (DSP) [46] constructs in PDSIDES. This includes the formulation of selection decisions that involve making a choice among a number of possibilities considering a number of measures of merit, and compromise decisions that determining the "right" values of design variables to describe the best satisficing solution with respect to constraints and multiple goals, using the selection DSP (sDSP) and the compromise DSP (cDSP) constructs, respectively. The combinations of these two types of decisions are formulated using the coupled DSP construct.
- ***Decision Workflows Formulation***. This functionality facilitates designers formulating decision workflows which are critical in meta-design of design processes using the PEI-X diagram [47]. Designers can use the entities defined in the so-called DSPT palette [47] to compose different decision workflows for representing design processes and analyzing them before they are implemented. This is important since it enables the exploration of process alternatives during the design stage.
- ***Solution Space Exploration***. This functionality facilitates designers exploring the solution space by following a formal procedure that centers on the cDSP construct. The procedure takes design requirements as input and gives satisficing design specification as output. Steps include clarification of the design event, definition of the problem, identification of theoretical and empirical models that are available, development of surrogate models, formulation of the cDSP, solving the cDSP and performing post-solution analysis. The solution space exploration functionality insures that the output design specification is acceptable.
- ***Uncertainty Management***. This functionality help designers deal with different types of uncertainties and identify robust solutions against the uncertainties using robust design indices such as the Design Capability Indices (DCI) [49] and the Error Margin Indices (EMI) [50]. With these functionalities, designers are able to make "safe" decisions that are relatively insensitive to variations.
- ***User/Activity Specific Decision Support***. This functionality tailors decision support to accommodate different types of users in various activities. In PDSIDES, we define three types of users for customized decision support, namely, *domain experts* who are responsible for creating decision templates in original design activities, *senior designers* who are responsible for editing (or tailoring) existing decision templates in adaptive design activities, and *novice designers* who are responsible for executing existing decision templates in variant design activities. This ensures that they each receive the appropriate decision support.

- **Connection to Other Platforms**. Since the basic structure for representing knowledge in PDSIDES is an ontology, in which the terms and relationships are explicitly and formally defined, this enables PDSIDES to exchange knowledge (particularly the decision templates) with other Product Lifecycle Management

(PLM) platforms such as product data management systems and simulation-based analysis systems.

In Table 1.1, we summarize the constituents of the architecture for PDSIDES. A similar table is included in each chapter to provide readers with an overview of what to expect in each chapter of the monograph.

In the chapters that follow we discuss how the functionalities of PDSIDES are realized using the constructs and techniques from Decision-Based Design and Knowledge-Based Systems, and how those functionalities are demonstrated using examples. In Fig. 1.6, we show how the functionalities of PDSIDES are mapped to different chapters where the key constructs for realizing the corresponding functionalities are discussed and the efficacy of the constructs are tested by using examples. The foundations of the architecture including the DSP constructs, PEI-X diagram, robust design framework, decision templates and ontologies, etc., are discussed in Chap. 2. Support for formulating individual decisions is covered in Chap. 3 where we develop templates and ontologies for selection, compromise, and hierarchical decisions. The support for formulating decision workflows is covered in Chap. 5 where an ontology based on the PEI-X diagram is developed. User-specific decision support is afforded in Chap. 4 where we present a prototype of PDSIDES to which different types of users and the corresponding working scenarios are defined and the associated requirements for decision support are discussed. Design space exploration and uncertainty management are covered in Chap. 6 where we present the procedure and the ontology for robust design space exploration under different types of uncertainties. Since PDSIDES is a knowledge-based platform, knowledge management is covered in most of the chapters, including Chaps. 3–6, where ontology development and knowledge reuse are involved. The logical flow of the chapters is discussed in Sect. 1.4.

Table 1.1 Summary of architecture for PDSIDES

Architecture constituents	Summary interpretation for a knowledge-based decision support platform in design of engineered systems
Elements	What?
Components	Key components of the decision support platform (refer to Fig. 1.5)
Connectors	The media that links the components, namely, the platform
Form	How?
Component roles	Different roles that the components play in the platform
Properties	Attributes or features of the components
Relationship	The relationships among different components in the platform
Rationale	Why?
Motivation	The underlying reason to use this platform, driven by requirements
Characteristics/beliefs	Characteristics, assumptions, and beliefs used to guide the selection of elements and form
Interpretation	Understanding the big picture of the decision support platform

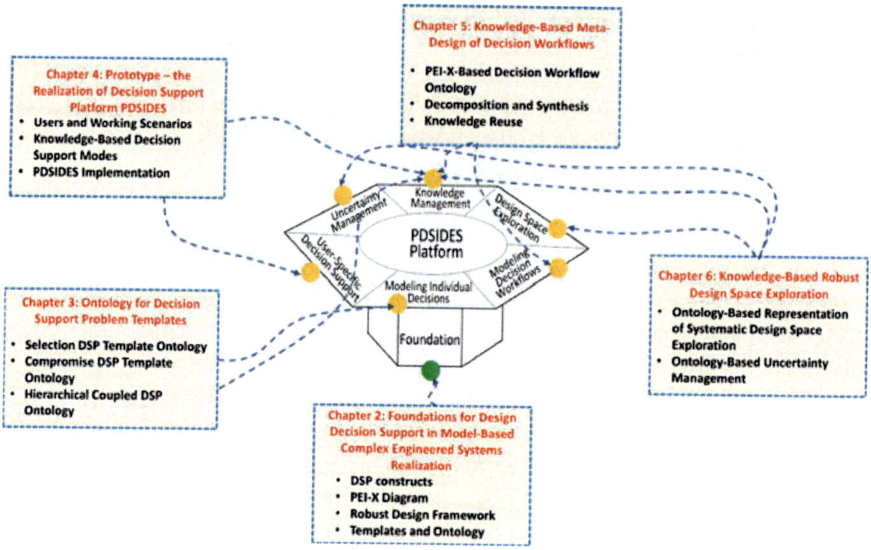

Fig. 1.6 Mapping the functionalities of PDSIDES to the chapters

1.4 Organization and Validation Strategy of the Monograph

The monograph is organized as a flowchart shown in Fig. 1.7. In this chapter, we introduce the background, motivation and the frame of reference, and propose the requirements and architecture for a decision support platform—PDSIDES. The foundational constructs and techniques for decision support in model-based realization of complex engineered systems are reviewed and discussed in Chap. 2. In Chap. 3, we develop the ontologies for representing the knowledge related to selection, compromise, and coupled decision templates, which can be used by designers to formulate individual decisions in design. Based on the decision templates and ontologies, we develop a prototype of PDSIDES in Chap. 4 which can support the three types of users making decisions in original, adaptive, and variant design. In Chaps. 5 and 6, we add more functionalities to PDSIDES so that it can provide knowledge-based support in meta-design of decision workflows and robust design space exploration under uncertainty. In Chap. 7 we summarize our contributions in this monograph, propose a conceptual framework for a Cloud-Based Platform for Decision Support in the Design of engineered Systems (CB-PDSIDES) and identify several open questions for future research, and provide some closing comments.

The relationship of research efforts with the constructs of the architecture and connection between chapters of the monograph are shown in Fig. 1.8. The ontologies for the selection and compromise decisions developed in Chap. 3 form the foundation for the ontology for hierarchical coupled decisions developed in the same chapter. And the three ontologies provide the knowledge representation schemes for

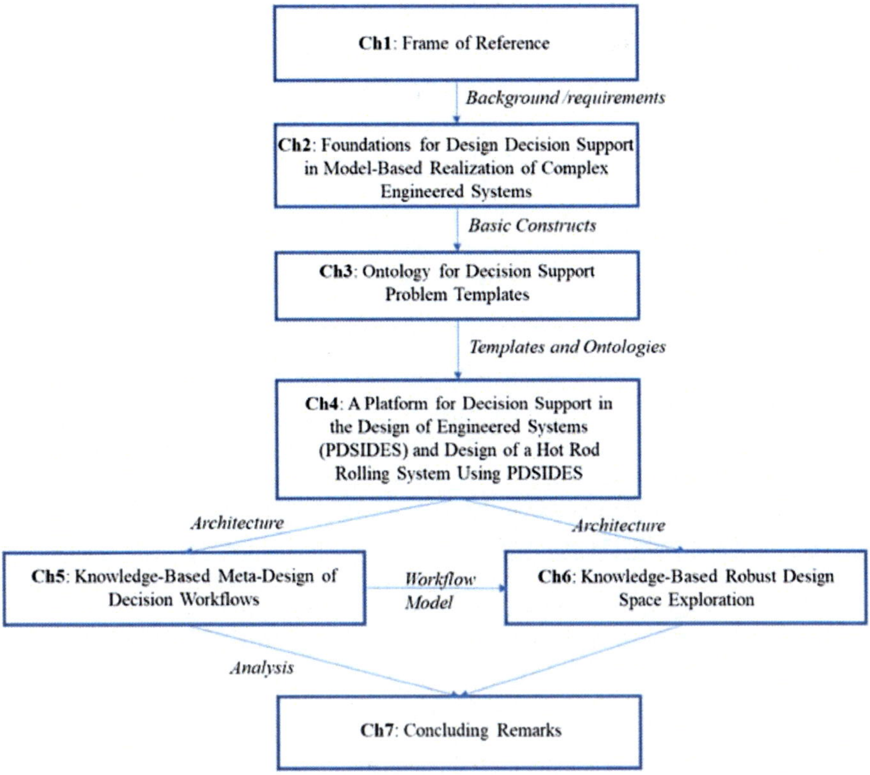

Fig. 1.7 The organization of the monograph

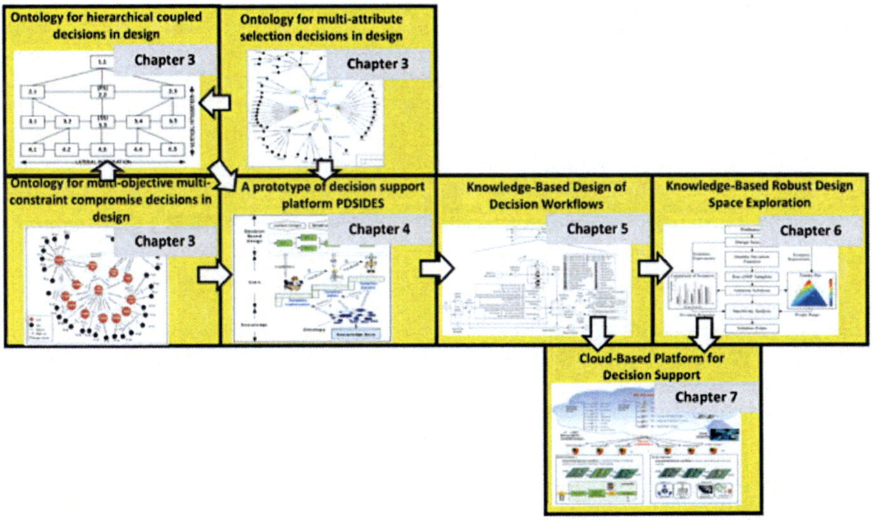

Fig. 1.8 Relationship of research efforts with the constructs of the architecture and connection between chapters of the monograph

the knowledge base of the prototype of PDSIDES developed in Chap. 4. Based on the prototype of PDSIDES, in Chap. 5 we extend the ontology to represent the knowledge of decision workflows based on the PEI-X diagram and enable the functionality of PDSIDES to support meta-design of decision workflows. Then, the idea of representing phase- and event-based procedural knowledge is extended from Chaps. 5 and 6 for representing a robust design space exploration processes, where the issues of uncertainty management and robust decision-making can be addressed in PDSIDES. Based on the current functionalities of PDSIDES, the focus of Chap. 7 is on furthering the research vision by exploring opportunities for a service-oriented architecture for decision support using the cloud.

The verification and validation strategy used in this monograph is based on the validation square proposed by Pederson et al. [51, 52]. Key to the validation square are the four quadrants: Theoretical Structural Validity (TSV), Empirical Structural Validity (ESV), Empirical Performance Validity (EPV), and Theoretical Performance Validity (TPV), as shown in the center square of Fig. 1.9. The corresponding verification involves checking for internal consistency. The goal of checking TSV is to verify the individual constructs constituting a method as well as the internal consistency of the integration of all constructs to form an overall method, therefore TSV consists of establishing requirements for the design method, carrying out a literature review, and establishing the logical soundness of constructs used—individually and integrated.

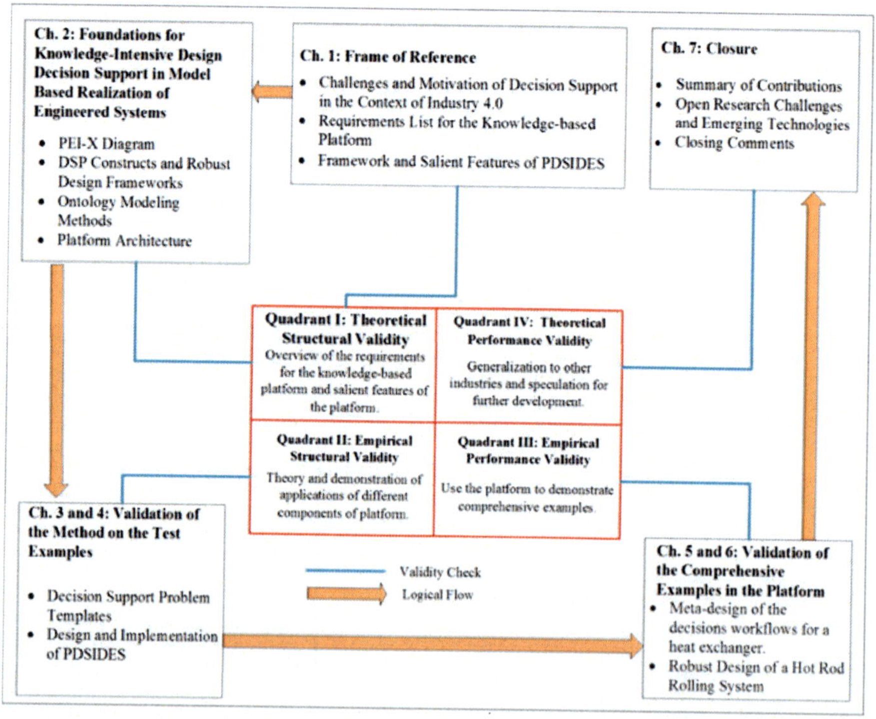

Fig. 1.9 The organization of the chapters according to the validation square

The goal of checking ESV is to build confidence in the appropriateness of the test example problems chosen for illustrating and verifying the performance of the design method, therefore ESV consists of checking the appropriateness of the test example problems selected to test the design method and accepting the design methods and constructs. The goal of checking EPV is to build confidence in the usefulness of the method using example problems, therefore EPV consists of checking the ability of the method to provide useful results for selected example problems. The goal of checking TPV is to build confidence in the generality of the design method and accepting that the method is useful beyond the example problems considered, so TPV consists of checking the ability to provide useful results beyond example problems and showcasing the generic form of method.

In this monograph, the mapping of the chapters to the validation square is shown in Fig. 1.9. In Chaps. 1 and 2 we cover a literature review, identify the requirements of a decision support platform, identify the components and the architecture of the platform, therefore they correspond to the TSV quadrant of the validation square where we check the structural and logical soundness of platform. In Chap. 3, we discuss the development of thee ontologies for representing the knowledge related to selection, compromise, and coupled decisions, respectively, and using three small examples (i.e., pressure vessel design, the light switch cover material selection, and portal frame design) to test their utility, therefore Chap. 3 maps to ESV where the appropriateness of the test examples are checked. In Chaps. 4–6, we highlight the development of a prototype platform PDSIDES and the functionalities of designing decision workflows. We describe robust design space exploration. We illustrate the efficacy using two examples, namely, a hot rod rolling system design problem and a heat exchanger design problem. In Chaps. 4–6, we map to Quadrant EPV and establish that the results from PDSIDES are useful. In Chap. 7, we summarize what is presented in the monograph and this maps to Quadrant TPV. We suggest that the platform as well as the associated constructs and functionalities are valid beyond the example problems selected for empirical validation. Further, in Chap. 7, we discuss avenues for future research and broader applications of the fundamental ideas in this monograph and propose the framework of cloud-based PDSIDES (CB-PDSIDES) and the associated research questions. Details associated with the particular verification and validation steps are discussed at the end of each chapter.

References

1. James, S., & Shetty, A. (2019). An initial framework for implementation of Industry 4.0 in the high technological manufacturing sector in Southern California. In *Proceedings ASME 2019 International Mechanical Engineering Congress and ExpositionV02BT02A012*.
2. Merkel, A. (2014). Speech by Federal Chancellor Angela Merkel to the OECD Conference. https://www.bundesregierung.de/breg-en/chancellor/speech-by-federal-chancellor-angela-merkel-to-the-oecd-conference-477432.
3. Xu, L. D., Xu., E. L., & Li, L. (2018). Industry 4.0: State of the art and future trends. *International Journal of Production Research,56*(8), 2941–2962.

4. Jiao, R., Commuri, S., Panchal, J., Milisavljevic-Syed, J., Allen, J. K., Mistree, F., & Schaefer, D. (2021). Design engineering in the age of Industry 4.0. *Journal of Mechanical Design, 143*(7), 070801.
5. Bagheri, B., Yang, S., Kao, H. A., & Lee, J. (2015). Cyber-physical systems architecture for self-aware machines in Industry 4.0 environment. *IFAC-PapersOnLine, 48*(3), 1622–1627.
6. Mistree, F., Smith, W. F., Bras, B. A., Allen, J. K., & Muster, D. (1990). Decision-based design: A contemporary paradigm for ship design. *Transactions of the Society of Naval Architects and Marine Engineers, 98*, 565–597.
7. ABET. (2017). Criteria for Accrediting Engineering Programs. https://www.abet.org/accreditation/accreditation-criteria/criteria-for-accrediting-engineering-programs-2016-2017/.
8. Zhou, F., Ji, Y., & Jiao, R. J. (2013). Affective and cognitive design for mass personalization: Status and prospect. *Journal of Intelligent Manufacturing, 24*(6), 1047–1069.
9. Zhou, J., Zhou, Y., Wang, B., & Zang, J. (2019). Human–cyber–physical systems (HCPSs) in the context of new-generation intelligent manufacturing. *Engineering, 5*(4), 624–636.
10. Kang, Y., Wang, W., Meng, L., Zheng, Y., Xiao, P., Song, F., and Cheng, X. (2019). From tool revolution to decision-making revolution—The road to intelligent manufacturing transformation. Report of KPMG and Ali Research.
11. Hazelrigg, G. A. (1996). *Systems engineering: An approach to information-based design.* Pearson College Division.
12. Hazelrigg, G. A. (1998). A framework for decision-based engineering design. *Journal of Mechanical Design, 120*(4), 653–658.
13. Lewis, K., & Mistree, F. (1998). Collaborative, sequential, and isolated decisions in design. *Journal of Mechanical Design, 120*(4), 643–652.
14. Lewis, K. E., Chen, W., & Schmidt, L. C. (2006). *Decision making in engineering design.* ASME Press.
15. Chen, W., Hoyle, C., & Wassenaar, H. J. (2013). Decision-based design. Proceedings. Springer.
16. Zhou, F., Xu, Q., & Jiao, R. J. (2011). Fundamentals of product ecosystem design for user experience. *Research in Engineering Design, 22*(1), 43–61.
17. Tucker, C., & Kim, H. (2011). Predicting emerging product design trend by mining publicly available customer review data. In *Proceedings DS 68-6: Proceedings of the 18th International Conference on Engineering Design (ICED 11), Impacting Society through Engineering Design* (Vol. 6). Design Information and Knowledge.
18. Jiao, R., & Zhou, F. Product line planning incorporating peer influence of social networks. In *Proceedings of IEEE International Conference on Industrial Engineering and Engineering Management, Thailand.*
19. Young, A. J., & Hargrove, L. J. (2015). A classification method for user-independent intent recognition for transfemoral amputees using powered lower limb prostheses. *IEEE Transactions on Neural Systems and Rehabilitation Engineering, 24*(2), 217–225.
20. Green, M. G., Tan, J., Linsey, J. S., Seepersad, C. C., & Wood, K. L. (2005). Effects of product usage context on consumer product preferences. In *Proceedings of ASME 2005 International Design Engineering Technical Conferences and Computers and Information in Engineering Conference* (pp. 171–185). American Society of Mechanical Engineers Digital Collection.
21. Zhang, W., Wang, S., Hou, L., & Jiao, R. J. (2021). Operating data-driven inverse design optimization for product usage personalization with an application to wheel loaders. *Journal of Industrial Information Integration, 23*, 100212.
22. Zhou, F., Jianxin Jiao, R., & Linsey, J. S. (2015). Latent customer needs elicitation by use case analogical reasoning from sentiment analysis of online product reviews. *Journal of Mechanical Design, 137*(7), 071401.
23. Zheng, P., Xu, X., & Chen, C.-H. (2020). A data-driven cyber-physical approach for personalised smart, connected product co-development in a cloud-based environment. *Journal of Intelligent Manufacturing, 31*(1), 3–18.
24. Hou, L., & Jiao, R. J. (2020). Data-informed inverse design by product usage information: A review, framework and outlook. *Journal of Intelligent Manufacturing, 31*(3), 529–552.

25. Baskarada, S., & Koronios, A. (2013). Data, information, knowledge, wisdom (DIKW): A semiotic theoretical and empirical exploration of the hierarchy and its quality dimension. *Australasian Journal of Information Systems, 18*(1), 5–24.
26. Zha, X. F., Sriram, R. D., Fernandez, M. G., & Mistree, F. (2008). Knowledge-intensive collaborative decision support for design processes: A hybrid decision support model and agent. *Computers in Industry, 59*(9), 905–922.
27. Rich, E. (1983). *Artificial intelligence.* McGraw-Hill.
28. Baxter, D., Gao, J., Case, K., Harding, J., Young, B., Cochrane, S., & Dani, S. (2008). A framework to integrate design knowledge reuse and requirements management in engineering design. *Robotics and Computer-Integrated Manufacturing, 24*(4), 585–593.
29. Formulation. https://en.wikipedia.org/wiki/Formulation.
30. Roth, B. M., & Mullen, J. D. (2002). *Decision making: Its logic and practice.* Rowman & Littlefield.
31. Kuppuraju, N., Ganesan, S., Mistree, F., & Sobieski, J. S. (1985). Hierarchical decision-making in system-design. *Engineering Optimization, 8*(3), 223–252.
32. Kang, E., Jackson, E., & Schulte, W. (2010). An approach for effective design space exploration. In *Proceedings of Monterey Workshop* (pp. 33–54). Springer
33. Glover, F., & Marti, R. (1997). *Tabu search.* Springer.
34. Mirjalili, S. (2019). Genetic algorithm. In *Evolutionary algorithms and neural networks* (pp. 43–55) Springer.
35. LaValle, S. M., & Kuffner, J. J., Jr. (2001). Randomized kinodynamic planning. *The International Journal of Robotics Research, 20*(5), 378–400.
36. Aarts, E. H. (1987). *Simulated annealing: Theory and applications.* Springer.
37. Vajda, S. (2009). Mathematical programming. Courier Corporation.
38. Lee, S. M. (1972). Goal programming for decision analysis. Auerbach Publishers.
39. Box, G. E. (1976). Science and statistics. *Journal of the American Statistical Association, 71*(356), 791–799.
40. Isukapalli, S., Roy, A., & Georgopoulos, P. (1998). Stochastic response surface methods (SRSMs) for uncertainty propagation: Application to environmental and biological systems. *Risk Analysis, 18*(3), 351–363.
41. Choi, H.-J., Mcdowell, D. L., Allen, J. K., & Mistree, F. (2008). An inductive design exploration method for hierarchical systems design under uncertainty. *Engineering Optimization, 40*(4), 287–307.
42. Oberkampf, W., DeLand, S., Rutherford, B., Diegert, K., & Alvin, K. (2012). A new methodology for the estimation of total uncertainty in computational simulation. In *Proceedings of 40th Structures, Structural Dynamics, and Materials Conference and Exhibit* (p. 1612).
43. Allen, J. K., Seepersad, C., Choi, H., & Mistree, F. (2006). Robust design for multiscale and multidisciplinary applications. *Journal of Mechanical Design, 128*(4), 832–843.
44. Simon, H. A. (1960). *The new science of management decision.* Harper and Row.
45. McKenney, J. L., & Keen, P. G. W. (1974). How managers minds work. *Harvard Business Review* (pp. 79–90).
46. Marston, M., Allen, J. K., & Mistree, F. (2000). The decision support problem technique: Integrating descriptive and normative approaches in decision based design. *Engineering Valuation & Cost Analysis, Special Issue on Decision-Based Design: Status and Promise, 3*(2), 107–129.
47. Mistree, F., Bras, B., Smith, W. F., & Allen, J. (1996). Modeling design processes: A conceptual, decision-based approach. *International Journal of Engineering Design and Automation, 1*(4), 209–221.
48. Gruber, T. R. (1993). A translation approach to portable ontology specifications. *Knowledge Acquisition, 5*(2), 199–220.
49. Chen, W., Simpson, T. W., Allen, J. K., & Mistree, F. (1999). Satisfying ranged sets of design requirements using design capability indices as metrics. *Engineering Optimization, 31*(5), 615–619.

50. Choi, H. J., Austin, R., Allen, J. K., Mcdowell, D. L., Mistree, F., & Benson, D. J. (2005). An approach for robust design of reactive power metal mixtures based on non-deterministic micro-scale shock simulation. *Journal of Computer-Aided Materials Design, 12*(1), 57–85.
51. Pedersen, K., Emblemsvåg, J., Bailey, R., Allen, J. K., & Mistree, F. (2000). Validating design methods and research: the validation square. In *Proceedings of the International Design Engineering Technical Conferences and Computers and Information in Engineering Conference,* Paper Number DETC2000/DTM-14579
52. Seepersad, C. C., Pedersen, K., Emblemsvåg, J., Bailey, R., Allen, J. K., & Mistree, F. (2006). The validation square: How does one verify and validate a design method. In *Decision making in engineering design* (pp. 303–314) American Society of Mechanical Engineers.

Chapter 2
Foundations for Design Decision Support in Model-Based Complex Engineered Systems Realization

In this chapter, we introduce the foundations for the decision support platform architecture proposed in Chap. 1. The foundations include (1) primary constructs in decision-based design, (2) frameworks for robust decision-making, (3) the PEI-X diagram for designing decision workflows, and (4) knowledge-based techniques for decision support. After introducing the foundations, we review the key elements in the decision support platform architecture and check theoretical structural validity. Table 2.1 is a summary of this chapter. The mapping of the sections to the components (topics) discussed in this chapter is presented in Row 2 of Table 2.1.

2.1 Primary Constructs in Decision-Based Design

Decision-Based Design (DBD) involves designing and creating design methods originating from the notion that decision-making is fundamental to design. We recognize that there are two types of implementation of DBD. One is the utility-based framework suggested by Hazelrigg [1], which exploits utility theory assuming objective rationality where designers seek to maximize the design utility using complete and accurate models. The second is the Decision Support Problem Technique (DSPT) by Mistree and co-authors [2] anchored in the notion of bounded rationality, proposed by Herbert A. Simon in the book *Sciences of the Artificial*, mandating satisficing solutions using models we know are incomplete, inaccurate, and not of equal fidelity. It is unrealistic to treat human decision-makers as objective rational beings. Herbert A. Simon proposed that there are restrictions in the capacity of human decision-makers to be fully rational and that we should treat them as "intendedly" rational, but only

© The Author(s), under exclusive license to Springer Nature Switzerland AG 2022

Z. Ming et al., *Architecting A Knowledge-Based Platform for Design Engineering 4.0*,
https://doi.org/10.1007/978-3-030-90521-7_2

Table 2.1 Summary of this chapter

Elements	What?
Components	Foundation to PDSIDES: DSP constructs (Sect. 2.1), robust design framework (Sect. 2.2), PEI-X diagram (Sect. 2.3), knowledge-based techniques (Sect. 2.4)
Connectors	Decision-based design
Form	How?
Component roles	Provides the foundational concepts to build the platform
Properties	Theoretical foundation
Relationship	Methods for modeling decisions, workflows, and knowledge
Rationale	Why?
Motivation	The need of supporting technologies for the platform
Assumptions	The principal role of a designer is to make decisions
Interpretation	Understanding the available technologies upon which the platform is going to be built

"boundedly" so. Simon outlines at least three ways in which actual behavior falls short of objective rationality [3]: (i) Rationality requires complete knowledge and anticipation of the consequences that will follow on each choice. In fact, knowledge of consequences in always fragmentary. (ii) Since these questions lie in the future, imagination must supply the lack of experienced feeling in attaching value to them. But values can only be imperfectly anticipated. (iii) Rationality requires a choice among all possible alternative behaviors. In actual behavior, only a very few of all these possible alternatives come to mind. In this monograph, we utilize the DSPT as our implementation of DBD. Key to DSPT is that there are two primary types of decisions in design, i.e., selection and compromise [4]. The selection Decision Support Problem (sDSP) [5] and the compromise Decision Support Problem (cDSP) [6] are the constructs for formulating the compromise and selection decisions, respectively. In Sects. 2.1.1 and 2.1.2, we discuss the sDSP and cDSP constructs in detail.

2.1.1 sDSP—The Selection Decision Support Problem

A selection decision in design is to indicate a preference for one among the potential alternatives based on multiple attributes. Selection is endemic throughout a design timeline. Decision Support Problems (DSP) afford the ability to model selection decisions, where several attributes are quantified utilizing insight-based *soft* and science-based *hard* information. Typically, the taxonomy of a design timeline information flow involves three categories: (1) the complete information utilized for the decision being *soft*, (2) part of the information being soft and the remaining being *hard*, and

Fig. 2.1 Word formulation for the preliminary selection DSP

Given	A set of concepts.
Identify	The principal *criteria* influencing selection The *relative importance* of criteria
Capture	Experience based knowledge about concepts with respect to a datum and established criteria
Rank	The concepts in order of preference based on multiple criteria and relative importance

(3) the entire information is *hard*. Based on the available information types, selection problems can be modeled as *preliminary selection* DSPs and *selection* DSPs. The former is formulated and solved if the decision exploits experience-based soft information, while the latter when significant hard information is available. *Preliminary selection* considers choosing the concept with the highest success probability that is further developed into viable alternatives. The DSP construct for representing *preliminary selection* is illustrated in Fig. 2.1.

Pugh's technique is the algorithm's basis to solve the *preliminary selection* DSP. In *preliminary selection*, choices are made based on some criteria to reduce the range of possible solution concepts to those with the highest success probability depending on the preferred performance of the solution [7, 8]. The *selection* DSP prioritizes the alternatives choices utilizing multiple features of varying importance, utilizing both experience-based subjective and science-based objective information. The priority order does not solely indicate the rank but also the quantitative preference of one choice over an alternative one. The DSP construct (word formulation and mathematic formulation) for representing *selection* is shown in Fig. 2.2.

It should be noted that for the two selection types, we use different terms for similar items. For the *preliminary selection,* we start with **concepts** and evaluate the concepts exploiting some **criteria** that are quantified using experienced-based judgment, and thus, a preliminary selection has to be utilized to define the top-of-the-heap concepts. Engineering is then applied to the possible top-of-the-heap concepts, and the **concepts** develop to **feasible alternatives**. During the *selection,* the **feasible alternatives** are evaluated based on **attributes** quantified using science-based data to identify the optimum alternative. For further information on the *selection* DSP, the reader is referred to [8], for its utilization in catalog design in [9, 10]. To deal with the performance uncertainty concerning attributes, the *selection* DSP is extended to the utility-based selection DSP (u-sDSP) [11] that combines utility theory and the sDSP structure. The u-sDSP formulation is illustrated in Fig. 2.3. The utility theory employed in the u-sDSP provides the mathematical rigor for decision-making under uncertainty, while the sDSP the construct and context to formulate and bound the decisions to facilitate practical use. The u-sDSP affords the designer a means to indicate preferences and identify the favored alternatives, which is more evident when uncertainty is considered in relation to alternative performance, i.e., variability. Additionally, the designer's role and his/her judgment are maintained to provide

***Selection* DSP – Word Formulation**	
Given	A set of *feasible alternatives.*
Identify	The principal *attributes* influencing selection The *relative importance* of attributes
Rate	The alternatives with respect to each attribute
Rank	The feasible alternatives *in order of preference* based on attributes and their relative importance

***Selection* DSP – Mathematic Formulation**	
Given	M *alternatives.* N *attributes* that influence the selection.
Determine	I_j *relative importance* of the jth attribute. Range 0 to 1. $A_{i,j}$ the *rating* of the ith alternative with respect to the jth attribute. $R_{i,j}$ the *normalized rating* of the ith alternative with respect to j^{th} attribute. Range 0 to 1.
Merit Function	$MF_i = \sum I_j R_{i,j}$ i = 1, ..., M (Range 0 to 1), j = 1, …, N
Select	Most promising alternative X_i^*: $X_i^* = max[MF_i]$

Fig. 2.2 Word and mathematic formulation of the selection DSP

Fig. 2.3 Utility-based selection DSP word formulation

Given	Finite set of *feasible alternatives.*
Identify	The principal *attributes* influencing selection. The *uncertainty* associated with each *attribute.*
Assess	Decision-maker's *utility* with respect to each *attribute* and with respect to *combinations of attributes.*
Evaluate	Each alternative using the decisionmaker's *utility functions.*
Rank	Most promising alternative(s) based on *expected utility.*

structure, context, critical evaluation, and justification during the decision-making procedure. The necessary stages to properly implement the u-sDSP are presented in [11], while the guidelines for assessing utility functions and calculating the expected utility are provided in Keeney and Raiffa [12].

2.1.2 cDSP—The Compromise Decision Support Problem

A compromise decision is to determine the proper values or their combination for the design variables (e.g., system parameters) to characterize the best satisficing system design concerning the considered constraints and goals [6]. Compromise decisions are formulated employing the cDSP construct illustrated in Fig. 2.4. A cDSP formulation is a hybrid formulation involving concepts from classic mathematical programming schemes and goal programming [13] combined with new concepts. It is closely related to goal programming in formulating multiple objectives or goals as system goals, and the objective function encompasses the goal deviation variables. Nevertheless, involving system constraints is retained from the classic constrained optimization formulation, as, unlike both other formulations, particular emphasis is given to the system's variable bounds. A detailed discussion about the difference between the cDSP, goal programming, and single-objective optimization is provided in [14] and thus is not repeated here.

The system descriptors are utilized to characterize a system's state completely. In this section, the cDSP's system descriptors are presented.

cDSP Word Formulation	*cDSP Mathematical Formulation*	
Given	**Given**	
Alternative to be improved	n	number of system variables
Assumptions used to model the domain	p+q	number of system constraints
Some helpful relations	p	equality constraint
The system parameter	q	inequality constraints
The constraints and goals for the system	m	number of system goals
	$g_i(X)$	system constraint function
		$G_i(X) = C_i(X) - D_i(X)$
	$f_k(X)$	function of deviation variables
Find	**Find**	
The values of the system variables	X_i	System variables $i = 1, \ldots, n$
Values of deviation variables	d_i^+, d_i^-	Deviation Variables $i = 1, \ldots, 2m$
Satisfy	**Satisfy**	
System constraints	*System constraints*	
	$g_i(X) = 0$	$i = 1, \ldots, p$
	$g_i(X) \geq 0$	$i = p+1, \ldots, p+q$
System goals (normalized)	*System goals (linear, nonlinear)*	
	$A_i(X) + d_i^- - d_i^+ = G_i$	$i = 1, \ldots, m$
Bounds on System Variables	*Bounds*	
	$X_j^{min} \leq X_j \leq X_j^{max}$	$i = 1, \ldots, n$
	$d_i^+, d_i^- \geq 0;\ d_i^+ \bullet d_i^- = 0$	
Minimize	**Minimize**	
	Deviation Function: Archimedean	
	$Z = \Sigma^m_{i=1} w_i (d_i^+, d_i^-)$	
Deviation Function	Deviation Function: Preemptive	
	$Z = [f_1(d_i^+, d_i^-), \ldots, f_k(d_i^+, d_i^-)]$	$i = 1, \ldots, m$

Fig. 2.4 Word formulation of the utility-based selection DSP

- **System variables** (X_i, $i = 1, \ldots, n$). The majority of the engineering scenarios involve a minimum of two system variables. Commonly, X represents n design variables that could be Boolean (0 for FALSE and 1 for TRUE), continuous, or their combination. By nature, system variables are unaffected by other descriptors, and the designer can change them as needed to modify the system's state. System variables employed to define an artifact are nonzero positive. In general, the n design variables within X represent the axis of an n-dimensional space.
- **System constraints** ($C_i(X) - D_i(X) \; =, \geq 0 \quad i = 1, \ldots, p + q$). These involve constraints that must be satisfied to assure a feasible design. Their mathematical notation involves functions that rely only on the system variables. System constraints are rigid, cannot be violated, and link the system's demand $D_i(X)$ with its capability $C_i(X)$ to attain the demand.
- **System goals** ($A_i(X) + d_i^- - d_i^+ \; = \; G_i \quad i = 1, \ldots, m$). These model the designer's aspiration and are expressed as equality, relating the designer's aspiration level G_i to the goal's true achievement $A_i(X)$. Nevertheless, the designer's aspiration might be quite high, or the system constraints too tight to manage the anticipated level of accomplishment.
- **Deviation variables** $\left(d_i^+, d_i^- \quad i = 1, \ldots, m\right)$. These permit the designer a specific latitude degree during the decision-making process. A specific goal could be overachieved $\left(d_i^- = 0 \text{ and } d_i^+ > 0\right)$ or underachieved $\left(d_i^- > 0 \text{ and } d_i^+ = 0\right)$. Therefore, the deviation variables link the true design performance with the aimed performance level and bound the requirement levels to be realistic. A deviation and a system variable differ in terms of the latter representing the distance from the design space origin in the ith dimension, while the former has as its origin the system's goal surface.
- **Bounds**. These are magnitude-related limits for each variable that involve an upper and lower bound resulting from the designer's judgment and the system's limitations. Most engineering design optimization methods neglect bounds despite these being necessary to restrict the system variables, i.e., $L \leq X \leq U$, defining the search space for a viable solution. If two or more system goals exist, all deviation variables must either be dimensionless or have the same dimension, preferably varying within a fixed range, for example, 0 to 1. To impose all deviation variables varying within a fixed range, it is mandatory to manipulate Gi's value correctly.
- **The objective**. In cDSP, the objective is minimizing the achievement function, $Z(d_i^+, d_i^-)$ expressed by exploiting the deviation variables. The aspiration level per goal is set by the designer, with the possibility of a design being impossible to meet the standards set. Therefore, the designer has to accept a compromise solution that affords a performance close to his aspirations. In essence, this is a compromise solution objective. To express the difference among goals and achievements, the deviation variables $Z(d_i^+, d_i^-)$ are used, which can be either arranged as an Archimedean function or a preemptive function (see [8] for details).

The solution algorithms are of two types, i.e., (1) approximately solving the exact problem and (2) solving a problem's approximation exactly. We solve the cDSPs utilizing the adaptive linear programming (ALP) [6] algorithm that belongs in the latter category. ALP relies on second-order terms regarding linearization, normaliation and transforming the constraints and goals into overall well-behaved convex functions within the region of interest, and employing a "smart" constraint accumulation and suppression strategy. Detailed steps for implementing ALP and a computational tool called "DSIDES" are used [15]. cDSP, ALP, and DSIDES are anchored in the satisficing (rather than optimizing) philosophy. Differences between optimization and satisficing strategy are summarized as follows [16].

(1) ***Differences between optimizing and satisficing strategy***

In summary, there are four advantages to using *satisficing* strategy which is realized by using cDSP, ALP and DSIDES. The advantages include the following.

The advantage in formulation

- Using Goals and Minimizing Deviation Variables Instead of Objectives.
  - The benefits are: At a solution point, only the necessary Kuhn-Tucker conditions are met, whereas the sufficient Kuhn-Tucker conditions do not have to be met.
  - Therefore, designers have a higher chance of finding a solution and a lower chance of missing a solution due to parameterizable and unparameterizable uncertainties.

The advantages in approximation:

- Using second-order sequential linearization
  - The benefit is: Designers can have a balance between linearization accuracy and computational complexity.
- Using accumulated linearization
  - The benefit is: Designers can manage non-convex problems, and deal with highly convex, nonlinear problems relatively more accurately.

The advantage in exploration:

- Combining interior-point searches and vertex searches:
  - The benefit is: Designers can avoid being stuck in local optimum to some extent and identify satisficing solutions which is relatively insensitive to the change in starting points.

The advantage in evaluation:

- Allowing some violations of soft requirements, such as the bounds of deviation variables.
  - The benefits are: Designers can manage rigid requirements and soft requirements in different ways to ensure feasibility.
  - As a result, goals and constraints with different scales can be managed.

(2) Differences among cDSP, Goal Programming, and Mathematical Programming

We summarize the differences among cDSP, Goal Programming, and Mathematical Programming seeking optimal solutions as follows.

Stage 1: Formulation. First, the cDSP is a hybrid of mathematical programming (seeking optimal solutions) and goal programming. In a cDSP, there are both constraints and goals. In goal programming, there are no constraints.[1] In a cDSP, the constraints are requirements (demands) that cannot be violated, whereas the goals are soft requirements (wishes) whose targets may not be reached but we want to minimize the distance between the targets and our results. The constraints and goals can be linear or nonlinear (convex or non-convex) or both, and equalities or inequalities or both. The benefit of being able to model both demands and wishes in one formulation is attractive in design. Due to the complexity of the supply chains for mass customization, the resulting cDSP often entails dealing with non-convex and convex constraints and goals.

Second, in a cDSP, we use deviation variables to assess the extent by which we over-achieve or under-achieve a goal. By adding the deviation variables, we ensure that the solutions to a cDSP satisfies the necessary Kuhn–Tucker condition. The solutions do meet the test of sufficiency to guarantee a "true" or "global" optimum. A *satisficing* solution to a cDSP is the mapping of an optimal solution on a lower-dimensional space. The dimensions being reduced consist of the deviation variables$D = [d_1^-, d_1^+, d_2^-, d_2^+, \ldots d_k^-, d_k^+]^T$. By adding deviation variables, we increase the dimensionality of a design problem, from $[x_1, x_2, \ldots x_n]^T$ to $[x_1, x_2, \ldots x_n, d_1^-, d_1^+, d_2^-, d_2^+, \ldots d_k^-, d_k^+]^T$, thus making it possible to absorb the risk of uncertainty at the constraint boundary. This results in a robust solution, that is, one that is relatively insensitive to uncertainties. By returning solutions consisting only of system variables, as is the case in solving an optimization problem,$X = [x_1, x_2, \ldots x_n]^T$, we decrease the dimensionality. Such "dimensionality reduction" may not result in a solution that is relatively insensitive to the uncertainties embodied in the modeling of the constraints of an optimization problem.

Third, although a cDSP has similarities with the auxiliary problem of a linear programming problem when using the two-phase method,[2] there are differences. For a linear programming problem (P), when we relax the m equality constraints $A \cdot X = b$ to m inequality constraints $A \cdot X + U = b$, by adding slack variables (or artificial variables) $U = [u_1, u_2, \ldots u_m]^T$, and change the objective function from $\min C^T \cdot X$ to $\min \sum_{i=1}^{m} u_i$, an auxiliary problem (A) of the original linear programming

[1] Although in later publications, the formulation of Goal Programming allows managing constraints. By the time Mistree and the coauthors published their work on cDSP and ALP [6], it was generally accepted that in Goal Programming, there are only "soft requirements" as goals but no "rigid requirements" as constraints.

[2] The introduction of the auxiliary problem is given at "http://www.math.uwaterloo.ca/~hwolkowi/henry/teaching/f05/350.f05/L18.pdf".

problem (P) is created. If a solution $[x_1, x_2, \ldots x_n, u_1, u_2, \ldots u_m]^T$ is optimal for (A) with $u_i = 0, i = 1, 2, \ldots m$, then the solution $[x_1, x_2, \ldots x_n]^T$ is feasible for (P).

There are similarities between a cDSP and an auxiliary problem (A). The slack variables U in the auxiliary problem are similar to the deviation variables D in a cDSP, if we treat the equality constraints $A \cdot X + U = b$ of (A) as the goals of a cDSP $\frac{Goal_i(X)}{Target_i} + d_i^- - d_i^+ = 1$. The objective of ($A$) is minimizing the sum of the slack variables U, similarly, the merit function of a cDSP is minimizing the linear combination of the deviation variables D.

There are differences, however, between a cDSP and an auxiliary problem (A) of a linear programming problem (P). An auxiliary problem is a linear problem, whereas a cDSP can be nonlinear—both constraints and goals. When solving an auxiliary problem (A), one can only obtain the feasibility of its original problem (P) but a *satisficing* solution (a good enough solution) is not guaranteed because the original objective function $\min C^T \cdot X$ is not incorporated in (A). On the contrary, in a cDSP, goals are satisfied as equality constraints in a corresponding optimization problem, thus, a *satisficing* solution that is close to achieve the goals is identified. In addition, in an auxiliary problem, for any constraint, we only minimize either its under-achievement or over-achievement, whereas in a cDSP, we can minimize both under-achievement and over-achievement of each goal. Furthermore, in an auxiliary problem, we treat all constraints equally by simply adding the slack variables U, while in a cDSP, we use weights to linearly combine the deviation variables so that we may assign different priorities to each goal.

In summary, a cDSP is different from goal programming, optimization, or an auxiliary problem in linear programming. Further a *satisficing* solution to a cDSP is not only feasible, but also adequate with respect to the achievement of the goals.

Stage 2: Exploration of the solution space. In the second stage, the solution space is explored to find *satisficing* solutions associated with each design preference (scenario), in different phases in the product life cycle. *Type I* and *Type II* uncertainty (refer to Sect. 2.2) are managed in the exploration of the solution space (ESS).

Weight sensitivity analysis—exploration of the design preferences. We use weight sensitivity analysis to explore how assessing different weights to the goals affects the system performance, that is, identifying satisficing solutions that are relatively insensitive to uncertainties.

System capacity analysis—identification and management of the sensitive segment and bottleneck. To overcome the capacity limitation of constraints or bounds, we propose system capacity analysis to identify the sensitive segment and bottleneck. If an inequality constraint has zero or a tiny surplus or slack compared with its right-hand side value, we define it as an active constraint. The solution is on or close to the boundary of the active constraint, so the solution is sensitive to the uncertainty of the active constraint. If the shadow price of an active constraint is lower than other active constraints, by relaxing this active constraint, we may not get much improvement in achieving the goals, and we define such a constraint as a "sensitive segment." We then move the solution away from the sensitive segment by restricting the active constraint, that is, by adding a buffer to the constraint to prevent the solution from reaching the boundary defined by the constraint. If the shadow price of an active constraint has the largest value in comparison with that of other constraints, relaxing the constraint can result in the greatest improvement of the achievement of the goals. We define such an active constraint as a "bottleneck." Also, it is important to find ways of relaxing the bottleneck in the physical system to boost the system potential. Once there is no longer the potential of physically relaxing the constraint, we move the solution away from the newly relaxed boundary by restricting the constraint by adding a buffer to the defined boundary. Thus, we balance the need for robustness of the solution with our desire to obtain the best satisficing solution.

2.2 Framework for Robust Decision-Making

In engineering design, the concept of robustness is used to alleviate the consequences of variability without removing its underlying sources [17–19]. The anticipated variability usually represents information originating from other parts of the product realization process that exceed the modeled and designed system's boundaries [17]. For example, a system may be designed for performance (response) that is relatively insensitive to design variations caused by the manufacturing processes. Characterization and classification of uncertainty are essential for developing robust design methods. In the literature concerning engineering design and analysis, uncertainty is presented employing several definitions and classifications. For example, Isukapalli et al. [20] suggest the following uncertainty types: (a) "natural uncertainty or variability" that involves the physical system's inherent randomness or unpredictability, (b) "model uncertainty" that it used to consider simplifications and approximations in model formulation, and (c) "data uncertainty" referring to the imperfect information regarding the model inputs or parameters. Based on Isukapalli's classification, Choi et al. [21] introduce the notion of propagated uncertainty caused by a chain of models in the context of hierarchical system design. Oberkampf et al. [22] distinguish uncertainty, variability, and error in computational simulations. In their definition, the definition of variability involves the inherent variation linked to a physical system, where uncertainty is the possible modeling process deficiency during any stage or activity due to the lack of knowledge, and error is defined as a recognizable modeling process deficiency during any stage or activity not linked to the absence of knowledge. Allen et al. [17] suggest from a system design perspective three uncertainty types: uncertainty in noise or due to environmental and other noise factors, uncertainty in design variables or control factors, and uncertainty introduced by modeling methods. These authors also suggest a robust design framework for multidiscipline and multiscale applications. In this monograph, we afford robust decision-making in engineering design by implementing the framework proposed by Allen et al. [17].

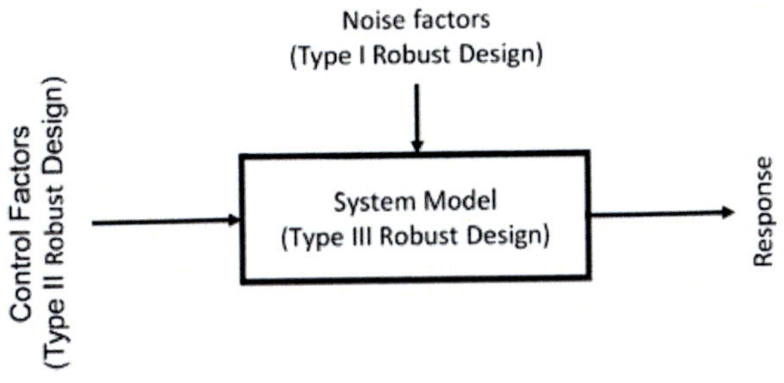

Fig. 2.5 Robust design types

According to Allen et al. [17], the three robust design types are (see Fig. 2.5):

- *Type-I robust design* is employed to determine the design variable (control factor) values satisfying a collection of performance requirement targets, regardless of the noise factor variations.
- *Type-II robust design* is utilized to determine the design variable values satisfying a collection of performance requirement targets, regardless of the control factor variations. For example, in the early design stages, the design variable values change as the design evolves; hence, it is preferable to identify starting values which, if changed, have the least possible impact on the system's performance and thus minimal iteration during the evolution of the design process.
- *Type-III robust design* is employed to determine the adjustable ranges of the design variables satisfying the performance requirements that are not sensitive to the system model variability. For example, a system model might include simplifying assumptions or random factors, e.g., random microstructure in materials design realizations, affecting the prediction's precision and accuracy.

Solutions to the three types of robust design can be depicted using the curves shown in Fig. 2.6. The solid bold line represents the nominal deviation or objective function (i.e., $Y = f(X)$). Due to system model uncertainties, the nominal deviation function may vary within the upper and lower limits represented with two dash lines. The optimal solution is the design variable X with a value that minimizes the nominal deviation function, denoted as A. When there is a variation (represented using a Gaussian distribution) around A, it can result in a dramatic variation in deviation Y, which means that system performance is susceptible to variation around the optimal solution. Robust design Types I and II seek to identify the curve's plateau where the deviation of the nominal function is relatively insensitive given the variations in both noise and control factors, as Solution B is shown in Fig. 2.6.

It should be noted that even though Solution B is not sensitive to control factors and noise variations, it is sensitive to system model variations. Solution C is the solution that is insensitive to control factors, noise variations, and the system model and is the solution for robust design Type I, II, and III. It can be seen that solution C is located at the curve's plateau and within the "narrowband" of the system model variation, and therefore solution C is a robust design solution that we look for in this monograph.

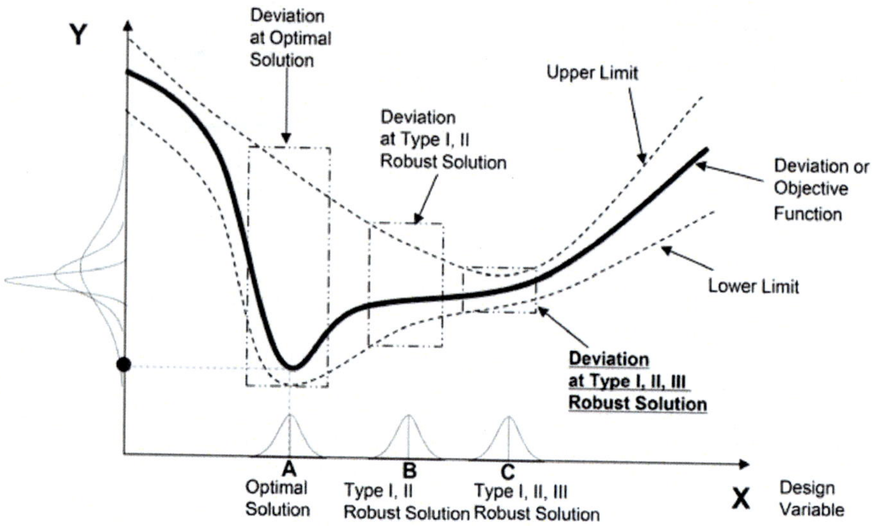

Fig. 2.6 Robust solutions in engineering design

In Fig. 2.7, we present the procedure to explore robust design solutions. The specific steps involved are as follows:

Step 0: Data Input/Output. The robust design solution exploration procedure starts with the designer identifying the system design requirements of the problem under

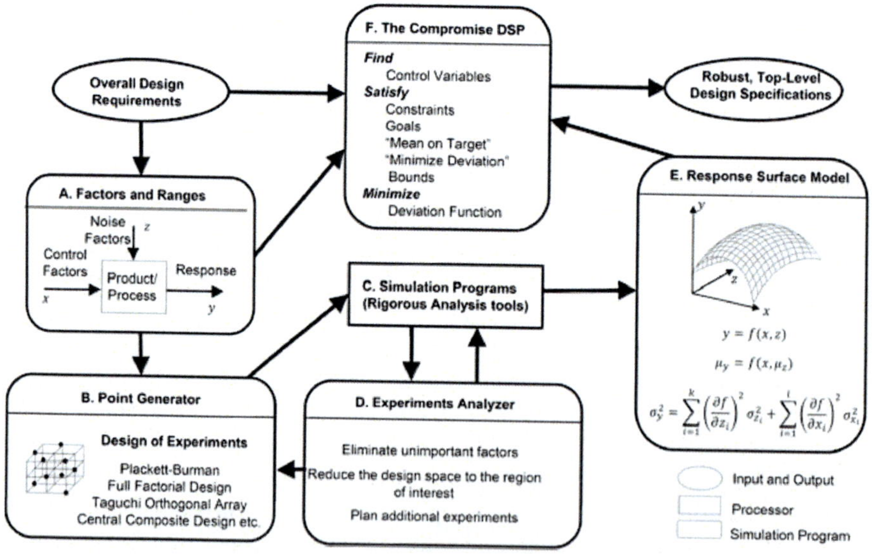

Fig. 2.7 Computing architecture of the robust concept exploration method [17]

study and concludes with determining the design parameters fulfilling the design requirements and satisfying the designer's wishes (goals).

Step 1: Pre-process—Processors A and F (Fig. 2.7). The designer identifies the factors influencing the system's performance and the ranges associated with formulating the cDSP. Control factors identified in Processor A will become the system variables in the cDSP formulation in Processor F, noise factors will become parameters of the cDSP, and ranges will become the bounds of the variables of the cDSP.

Step 2: Design of Experiments and Simulation—Processors B, C, and D (Fig. 2.7). The designer establishes the experiments to generate sample points in the factor space, run simulations, and analyze the simulation results to iteratively refine the experiments. Examples of design experiments include the Central Composite, Plackett–Burman, Taguchi Orthogonal Arays, and Full Factorial. The generated sample points are sent to simulation programs for computing the responses. The analyzer (Processor D) analyzes the simulation results to remove factors that are not important, restrict the design within the region of interest, and plan supplementary trials.

Step 3: Response Surface Modeling—Processor E in Fig. 2.7. The sample points and their associated responses are exploited to construct the response surface model for function $y = f(x, z)$. The response surface models are the approximate models for the constraints and/or goal functions in the cDSP formulation in Processor F.

Step 4: Compromise DSP formulation for Robust Design—Processor F in Fig. 2.7. In this step, robust design indices such as Error Margin Indices (EMI) [23] and Design Capability Indices (DCI) [24] are incorporated as goals in the cDSP formulation for bringing the "mean on the target" as well as "minimizing the deviation". The DCI is a collection of metrics specifically designed to assess the capability of a ranged set of design conditions whether it satisfies some design requirements. DCI addresses robust design Types I and II. EMI is a mathematical model highlighting the system's average performance location and the spread regarding the variability in both design variables and system models. EMI is utilized to address robust design Type III. For more information regarding combining EMI-DCI in a cDSP to address robust design Types I, II, and III, the reader is referred to [25].

2.3 Utilizing PEI-X Diagrams to Design Decision Workflows

From the viewpoint of DBD, design processes can be modeled decision workflows comprised of a series of decisions. In our DSPT framework, the DSPT palette [26] provides the entities for modeling decision workflow. The entities are used to construct hierarchies and model decision workflows regardless of the application

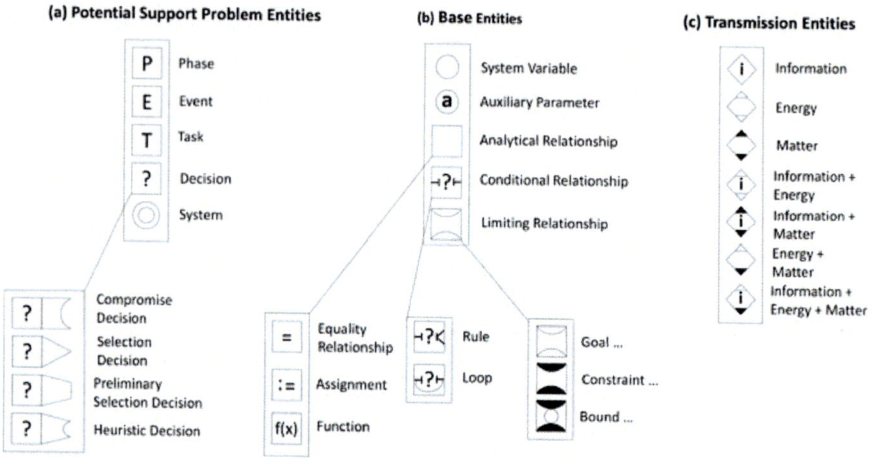

Fig. 2.8 DSPT palette entities (modified from Ref. [26])

domain. DSPT involves the distinct entity classes: base, potential support problem, and transmission entities, as shown in Fig. 2.8.

Potential support problem entities. P identifies the phase icon and represents the parts of a split process. Events occur within a phase, and E is used for identifying the event icon. Systems and/or human designers have direct involvement in tasks and decisions. On the other hand, performing tasks (i.e., an activity to be accomplished) and making decisions manage to accomplish phases and events. For the design team, the design process itself is a task containing other tasks and decisions, or even phases and events.

Nevertheless, trivial tasks such as "running computer program A" do not contain any decisions. T identifies a task and a rectangle with a question mark within it defines the decision icon in the palette. Concurrently, we include selection, preliminary selection, compromise, and heuristic decisions (not primary) [3]. A system is defined as a circle within a smaller circle.

Base entities. These involve basic objects to model design procedures which are computer-implemented and/or easily understood by designers. System variables are embedded in systems. Auxiliary parameters are utilized to model a process but are not system variables, e.g., counters in the loop. Although auxiliary parameters can be multidimensional, system variables are always scalars. Relationships are represented as rectangles, and thus rectangular-shaped icons are relationships. Phase, tasks, events, and decisions are also relationships. Analytical relationships are divided into equality relationships, assignments, and functions. Conditional relationships are rules and loops. Icons consisting of a nozzle within a rectangle indicate limiting relationships. Limiting relationships are grouped into goals, constraints, and bounds.

Transmission entities. Most of these entities rely on input and provide an output. Transmission entities are used to achieve connections and capture the input and output. Transmission entities involve three types: information, energy, and matter [27, 28], while their combinations may also occur. A transmission object commonly includes a catalog of other DSPT palette objects transmitted from one object to another. For example, the task "Provide the goals and constraints" outputs an information transmission object embedding a catalog of constraint and goal objects.

The Phase-Event-Information-X (PEI-X) diagram models the timeline and the hierarchy of decision-based design processes exploiting the DSPT palette entities. In such a diagram, the X may be Decisions, Tasks, and Systems. It ultimately provides a way to represent the life cycle timeline hierarchically. In Fig. 2.9 we show a part of a frigate's life cycle timeline, including the design phases, events, and product-specific information, highlighting the increasing qualitative relationship among hard and soft information. In this example, the design process is split into four major design stages on the top line. Commonly, each phase does not end abruptly, and thus usually, it is not trivial to define the start of a new stage. The overlapping stages are presented in Fig. 2.9. Several events can be identified in these phases. The horizontal bar in Fig. 2.9 provides an indication of the duration, in physical time, of phases and events.

The design procedure inputs a strategic need or foreign policy, and the whole process is represented in terms of phases, events, and information. Accordingly, designing involves generating the product-specific information identified as "Strategic Need/Foreign Policy" and ends with "Service Life History". In a desktop

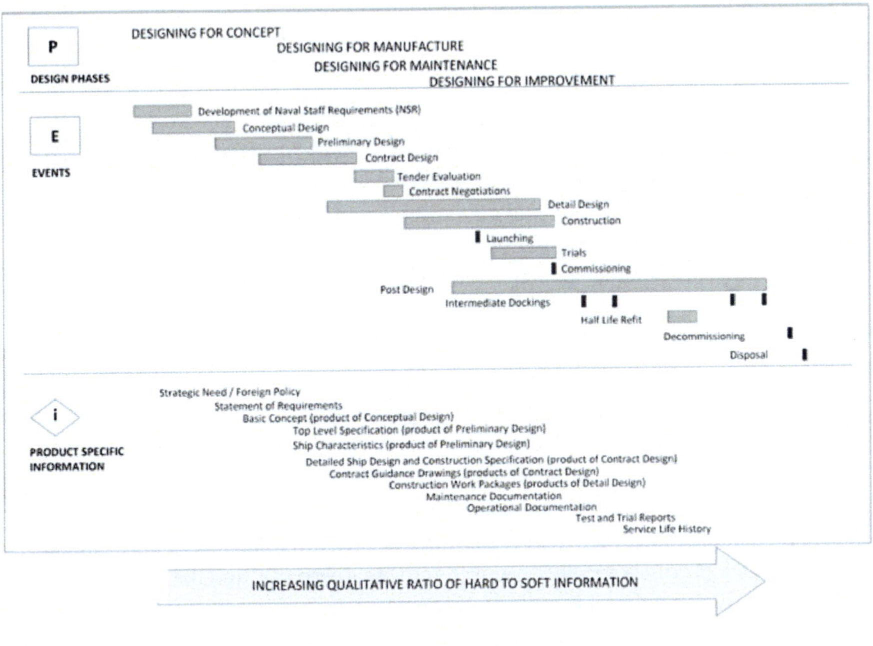

Fig. 2.9 PEI-X diagram for designing a frigate (modified from Ref. [3])

environment, a designer can "open" the lower-level model included in a specific object by clicking its icon. The lower-level models of the phase icons are represented using a network of the DSPT palette entities. An example of the designing phase icon for the concepts is illustrated in [26].

2.4 Knowledge-Based Techniques for Decision Support

Decision-making is a knowledge-intensive procedure, where knowledge is vital both in accelerating and affecting the decision procedure. Providing decision support from a knowledge-based perspective is critically important for enhancing the role of the designers as decision-makers. In Sect. 2.4, we introduce three techniques that we use in this monograph for knowledge-based decision support: template-based knowledge capture and reuse, ontology-based knowledge formalization, and knowledge-based platformization.

2.4.1 *Template-Based Knowledge Capture and Reuse*

Design processes can be described employing an information transforming equation [29] that includes an input, output, and a function that converts the input to the output, as shown in Eq. 2.1:

$$K = T(I) \tag{2.1}$$

where I denotes the requirement information (e.g., design goals, constraints, parameters), K the knowledge of product or system specifications after the design (e.g., design variable values, design goal deviations), and T the design process (e.g., formulating the design problem employing the cDSP construct and solving it using the ALP algorithm). In a computational environment, implementing the design process equation (Eq. 2.1) in a reusable manner is very important because process reusability can bring great value. A conservative estimation [30] suggests that more than 75% of design activity consists of case-based design—reuse of previous design knowledge to address a new design problem. In the light of this fact, it is necessary to capture and store the knowledge related to previous design processes so that designers can reuse it for solving similar or new design problems which will save them a lot of time in design from scratch. The idea of reusability has a long history in the industry. A typical example is to achieve reusability through modularization and standardization. Complex products are decomposed into smaller parts, and the ones with similar functions are grouped and standardized so that they can be reused in the assembly of different products, reducing the time and cost of developing a new product. In this work, we extend the concept of reusability to model design processes and

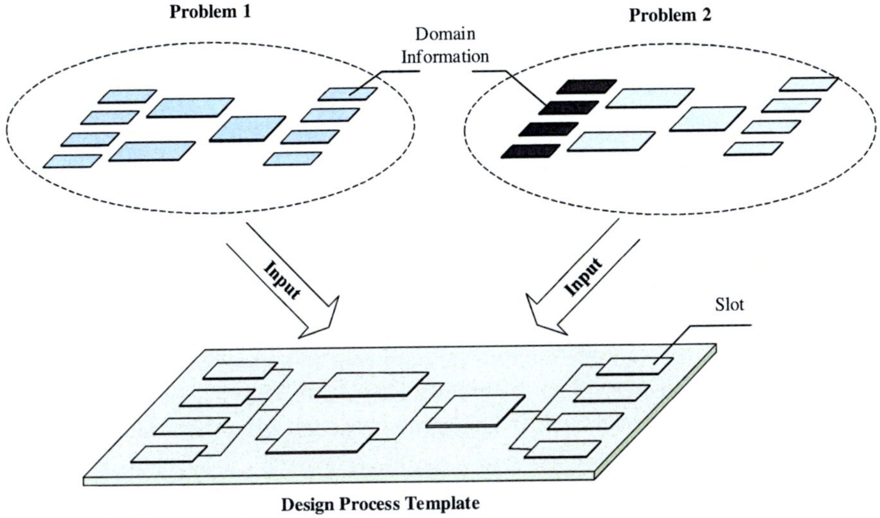

Fig. 2.10 Modular template for knowledge capture and reuse (modified from Ref. [32])

propose reusable and executable templates [31, 32], as shown in Fig. 2.10. A process template consists of two types of components, namely, a "bread board" with several connected slots, which stands for the procedural knowledge of the design process, and a "chipset", which stands for the declarative knowledge and can be installed onto the "bread board". The "bread board" represents the domain-independent know-how (e.g., the DSP constructs) that are constant across problems, and the "chipset" represents the domain-specific know-what (e.g., the variables, parameters, constraints, goals in a cDSP and alternatives and attributes in an sDSP) that varies from problem to problem. By separating the declarative and procedural knowledge, the reusability of both the "bread board" and "chipset" can be achieved. The "bread board" can be reused to instantiate any design process of the defined structure, and the "chipset" for instantiating processes with similar elements.

In PDSIDES, decision constructs (including the sDSP, cDSP, and coupled DSP) and the workflows composed using the decision constructs are represented as templates to facilitate the capture and reuse of decision-related knowledge. Three different types of template users are defined in PDSIDES, are discussed in Sect. 2.4.3.

2.4.2 *Ontology-Based Knowledge Formalization*

Knowledge related to the DSP constructs, design processes, and templates must be formalized to facilitate managing, retrieving, and sharing on PDSIDES. Ontologies, defined as explicit formal specifications of terms and their relations [33], are gaining attention for knowledge management in engineering. The knowledge model of Protégé, which consists of facets, classes, instances, and slots [34], is widely used for constructing ontologies. The critical components of the Protégé model are discussed as follows.

- **Classes and Instances**. A class is an object collection, where an object is a class *instance*, and the class is the instance's type. Classes follow a taxonomic hierarchy, e.g., if class `A` is B's subclass, then each instance of `A` is also of `B`. Protégé's class hierarchy root is the built-in class `THING`. A class may have multiple super-classes, enabling modeling various concepts' classifications in a domain [35].
- **Slots**. Slots describe the entity properties in the domain, e.g., a slot `hasUnit` can describe the variable's unit. Slots are frames and exist regardless of their attachment to classes. When a top-level slot is attached to a class, it becomes a template slot of the class, which is inherited to subclasses. A *slot* can have a *value* when the class is instantiated.
- **Facets**. Facets define the slot's properties and the constraints on the allowed slot values, which involve the maximum and minimum value for a numerical slot, cardinality (e.g., multiple, single), and type of value (e.g., Instance, String, Boolean, float).
- **Meta-classes**. A meta-class is a class whose instances are classes themselves and is a template to build classes, similarly to classes being templates to build instances. Classes are also meta-classes instances. Meta-classes are used to define the properties of the classes themselves, rather than properties of the instances of the classes.
- **Constraints, Axioms, and Rules**. Although the protégé knowledge model is quite complete, limitations to what can be expressed using classes, instances, slots, facets, and meta-classes still exist. Languages such as the Protégé Axiom Language (PAL)[3] and Java Expert System Shell (JESS)[4] are complementary to the Protégé knowledge model for representing complex constraints, axioms, and rules.

There are several methods for developing ontologies, i.e., the IDEF5 method proposed by Peraketh et al. [36] and the 7-step method proposed by Noy et al. [37]. The development of ontologies is an iterative process and there is no "correct" way to implement them, as one needs to keep revising and refining to obtain a "good enough" ontology. In this monograph, we use the 7-step method [37] for developing decision and workflow ontologies for PDSIDES since the method relies on the Protégé knowledge model. The seven steps are:

Step 1. **Determine the ontology's domain and scope**. The ontology process initiates by setting its domain and scope by answering questions such as, what is the ontology's domain coverage? What is the ontology's scope? To what types of questions should ontology information provide answers?

[3] https://protege.stanford.edu/plugins/paltabs/pal-documentation/overview.htm.

[4] https://jess.sandia.gov/

Who will employ and maintain the ontology? The goal in PDSIDES is to provide decision support during the design of the engineered systems, and the ontologies are used in PDSIDES to model the knowledge linked to different decisions types and workflows in a reusable and executable manner. Therefore, the domain of the ontologies for PDSIDES is engineering design, and the scope is the decision-based design where decisions are the core of design processes.

Step 2. **Consider reusing existing ontologies**. Reusing (refining or extending) current ontologies saves time in building new ontologies and facilitates the interaction with other applications or systems where some existing ontologies are embedded. In PDSIDES, we reuse the controlled vocabularies in describing the sDSP, cDSP, and the PEI-X diagram to communicate with the computational tool DSIDES [15].

Step 3. **Enumerate important terms in the ontology**. Obtaining a complete catalog of terms is necessary for describing the domain of interest. In the context of PDSIDES, these terms include *Design, Decision, Compromise, Selection, Decision Support Problem*, and *Workflow*, etc. At this step, there is no concern of overlapping concepts or properties these might have, links between the terms, or if concepts are slots or classes. Steps 4 and 5 deal with these problems by defining the classes and the class hierarchy, and the properties (slots) of classes. Steps 4 and 5 are closely related, and thus it is not trivial to apply them in a specific order.

Step 4. **Define the classes and the class hierarchy**. Developing a class hierarchy involves three potential strategies [38]: top-down, bottom-up, and *combination* development processes. The first starts by defining the most general concepts within the domain and then refines the concepts. The following strategy defines the most specific classes. The hierarchy is then removed and then these classes are grouped into general concepts. The last strategy combines the two former ones. In the context of PDSIDES, we adopt the bottom-up approach. We start with defining specific individual decisions and then defining a more general ontology for the PEI-X diagram of decision workflows. Details of the classes hierarchy are presented in Chaps. 3, 5, and 6.

Step 5. **Define the properties of classes—slots**. Once the class definition process has ended, we must describe the internal structure (i.e., the properties) of the concepts. In the Protégé knowledge model, properties are defined separately from classes. When a property is linked to a class to describe it, the class becomes a slot. In an ontology, several types of object properties can become slots, i.e., (1) "intrinsic" properties, e.g., a constraint's *linearity*, (2) "extrinsic" properties, e.g., a constraint's *name*; (3) parts, for a structured object, e.g., the *parameters* or *coefficients* in a constraint, (4) relationships to other individuals, e.g., the *creator* of a constraint.

Step 6. **Define the slots' facets**. A slot may have various facets to describe several value aspects such as type, the values allowed, cardinality, and other value characteristics. For instance, the value of a *unit* slot, e.g., "the unit of a variable", involves one string, while the *hasVariable* slot, e.g., "a decision has these variables", may include multiple values, which are *Variable* class instances.

Step 7. **Create instances**. In the last stage in the hierarchy individual class instances are created. To define an individual class instance involves (1) selecting a class, (2) creating an individual instance of that class, and (3) completing the slot values. For instance, an individual *Length* is created to characterize a specific *Variable* instance, having the following slot values: [Name: Length; LowerBound: 5; Upperbound: 100; Unit: inch; Symbol: l; QuantityType: Variable; Description: the length of a pressure vessel].

2.4.3 *Knowledge-Based Platform for Decision Support*

In the literature, many knowledge-based systems have been developed to support engineering design. For example, Shah et al. [39] introduce a parametric and feature-based system that quickly embeds specific algorithms and data structures to generate a reusable 3D geometric model. However, the parametric feature-based knowledge representations of [39] do not directly represent a human decision-making procedure during the design. Coyne et al. [40] suggest a prototype-centric scheme to develop knowledge-based design systems, where prototypes are generated, refined, and adapted to develop novel designs. Nevertheless, this work does not address the design decision-making processes. Finger and Dixon [41] surveyed several prescriptive, descriptive, and computer-based models of design processes in the late 1980s aiming to develop intelligent CAD expert systems. However, emphasis and analysis of the human decision-making procedure are inadequate, and the "concept selection" procedure is considered superficial without presenting any detailed information. Verhagen et al. [42] analyze 50 studies related to knowledge-based engineering (KBE), highlights the challenges, and proposes future research directions. Nevertheless, the authors state that all KBE methods examined automate the product design and development procedure without supporting designers to improve their decisions. Rocca [43] extensively reviews KBE regarding their language-based technological aspect to realize technological fundamentals and their utilization to automate large parts of the design process. This paper employs KBE to develop MDO multi-model generators, but the compromise decision (i.e., the tradeoff) among multidisciplinary models is not discussed. Jakiela and Papalambros [44] introduce a prototype "intelligent" CAD system, where the decision-making procedure during the conceptual design automatically utilizes production rules to develop 3D models. Although this method affords knowledge-based automatic decision-making during the design stage, it only considers geometrical modeling. Sapuan [45] presents a knowledge-based setup that is appropriate for material selection. Nevertheless, the decision procedure and the related knowledge representation language are linked to an explicit domain and are inextensible and non-reusable.

Although several knowledge-based setups exist, supporting the decision workflow during complex engineering systems design is still a challenge which is not well addressed primarily due to: (i) unavailability of executable and reusable decision knowledge representation methods. Indeed, knowledge reusability is essential during an adaptive and variant design, as the bulk of the initial decision workflow is preserved, i.e., reusable, and only minor part changes. (ii) Lack of user classifications to support decisions. The requirement of designers to support decisions varies depending on the design novelty involved and the designers' knowledge of the design process. Very few knowledge-based systems can recognize the various users' requirements and provide practical decision support.

To overcome the challenge of supporting the decision workflow in the design of complex engineering systems and address the requirements mentioned earlier, in the PDSIDES platform, we incorporate the decision workflow templates and the associated ontologies and classify users into three template categories, i.e., *template editors, template creators,* and *template implementers,* for tailored decision support. Details of the architecture and implementation of PDSIDES are discussed in Chap. 5.

2.5 Theoretical Structure Validity

According to the Validation Square [46], Theoretical Structural Verification and Validation (TSV) are to accept the individual constructs constituting a method as well as the internal consistency of the integration of all constructs to form an overall method. Theoretical structural validation involves systematically identifying the scope of the proposed approach's application, reviewing relevant literature and identifying the research gaps that exist, identifying the strengths and limitations of the constructs uses based on literature review, determining the constructs and approaches that can be leveraged for architecting PDSIDES while reviewing literature on the advantages, disadvantages and accepted domains of application, and checking the internal consistency of the constructs both individually and when integrated.

In this chapter, we establish the foundational constructs and approaches for building a knowledge-based platform for decision support in the design of engineered systems. We also justify why these constructs and approaches (including sDSP, cDSP, robust design framework, PEI diagram, decision templates, ontology) are appropriate for architecting the PDSIDES platform. These constructs and approaches have been previously applied successfully for problems in various domains and have been verified and validated. The use of these generic constructs and approaches for architecting a platform for decision support in the design of engineered systems is not addressed in past literature. The theoretical structural validity of PDSIDES is accepted by the logic procedural of literature review, gap analysis, and the evaluation of individual and integrated constructs. Empirical studies need to be carried out to establish the usefulness and effectiveness of PDSDIES and its associated constructs, which is addressed in Chaps. 3, 4, 5, and 6.

2.6 Where We Are and What Comes Next?

In this chapter we enumerate the foundations including DSP constructs, robust design framework, PEI-X diagram, decision template, and ontology for architecting the PDSIDES platform, the summary in terms of elements, form, and rationale is presented in Table 2.1. In the next chapter, we will develop ontologies for DSP templates which form the core for providing knowledge support in individual decision-making.

References

1. Hazelrigg, G. A. (1998). A framework for decision-based engineering design. *Journal of Mechanical Design, 120*(4), 653–658.
2. Mistree, F., Smith, W. F., Bras, B. A., Allen, J. K., & Muster, D. (1990). Decision-based design: A contemporary paradigm for ship design. *Transactions of the Society of Naval Architects and Marine Engineers, 98*, 565–597.
3. Simon, H. A. (1947). *Administrative Behavior*. Macmillan.
4. Mistree, F., Smith, W. F., Kamal, S. Z., & Bras, B. A. (1991). Designing Decisions: Axioms, Models and Marine Applications. In *Fourth International Marine Systems Design Conference, Kobe, Japan* (pp. 1–24).
5. Mistree, F., Lewis, K., & Stonis, L. (1994). Selection in the conceptual design of aircraft. *AIAA/NASA/USAF/ISSMO symposium on multidisciplinary analysis and optimization* (pp. 1153–1166).
6. Mistree, F., Hughes, O. F., & Bras, B. A. (1993). The compromise decision support problem and the adaptive linear programming algorithm. In M. Kamat (Ed.), *Structural optimization: Status and promise* (pp. 247–286). AIAA.
7. Kuppuraju, N., Ittimakin, P., & Mistree, F. (1985). Design through selection: A method that works. *Design Studies, 6*(2), 91–106.
8. Mistree, F., Marinopoulos, S., Jackson, D. M., & Shupe, J. A. (1988). The design of aircraft using the decision support problem technique. National Aeronautics and Space Administration, Scientific and Technical Information Division, NASA Contractor Report 4134.
9. Bascaran, E., Bannerot, R. B., & Mistree, F. (1989). Hierarchical selection decision support problems in conceptual design. *Engineering Optimization, 14*(3), 207–238.
10. Vadde, S., Allen, J. K., & Mistree, F. (1995). Catalog design—Selection using available assets. *Engineering Optimization, 25*(1), 45–64.
11. Fernandez, M. G., Seepersad, C. C., Rosen, D. W., Allen, J. K., & Mistree, F. (2005). Decision support in concurrent engineering—The utility-based selection decision support problem. *Concurrent Engineering Research A, 13*(1), 13–27.
12. Keeney, R. L., & Raiffa, H. (1976). *Decisions with multiple objectives : Preferences and value tradeoffs*. Wiley.
13. Ignizio, J. P. (1985). Multiobjective mathematical programming via the MULTIPLEX model and algorithm. *European Journal of Operational Research, 22*(3), 338–346.
14. Mistree, F., & Allen, J. K. (1997). Position paper optimization in decision-based design. In *Optimization in industry*. Palm Coast.
15. Reddy, R., Smith, W., Mistree, F., Bras, B., Chen, W., Malhotra, A., Badhrinath, K., Lautenschlager, U., Pakala, R., & Vadde, S. (1996). *DSIDES User Manual*. Systems Realization Laboratory, Woodruff School of Mechanical Engineering, Georgia Institue of Technology.
16. Guo, L. (2021). *Model evolution for the realization of complex systems*. Ph. D. Doctorial dissertation, The University of Oklahoma.

17. Allen, J. K., Seepersad, C., Choi, H., & Mistree, F. (2006). Robust design for multiscale and multidisciplinary applications. *Journal of Mechanical Design, 128*(4), 832–843.
18. Nair, V. N., Abraham, B., MacKay, J., Box, G., Kacker, R. N., Lorenzen, T. J., Lucas, J. M., Myers, R. H., Vining, G. G., & Nelder, J. A. (1992). Taguchi's parameter design: A panel discussion. *Technometrics, 34*(2), 127–161.
19. Tsui, K.-L. (1992). An overview of Taguchi method and newly developed statistical methods for robust design. *IIE Transactions, 24*(5), 44–57.
20. Isukapalli, S., Roy, A., & Georgopoulos, P. (1998). Stochastic response surface methods (SRSMs) for uncertainty propagation: Application to environmental and biological systems. *Risk Analysis, 18*(3), 351–363.
21. Choi, H.-J., McDowell, D. L., Allen, J. K., & Mistree, F. (2008). An inductive design exploration method for hierarchical systems design under uncertainty. *Engineering Optimization, 40*(4), 287–307.
22. Oberkampf, W., DeLand, S., Rutherford, B., Diegert, K., & Alvin, K. (1999). A new methodology for the estimation of total uncertainty in computational simulation. In *Proceedings of 40th Structures, Structural Dynamics, and Materials Conference and Exhibit* (p. 1612).
23. Choi, H. J., Austin, R., Allen, J. K., Mcdowell, D. L., Mistree, F., & Benson, D. J. (2005). An approach for robust design of reactive power metal mixtures based on non-deterministic micro-scale shock simulation. *Journal of Computer-Aided Materials Design, 12*(1), 57–85.
24. Chen, W., Simpson, T. W., Allen, J. K., & Mistree, F. (1999). Satisfying ranged sets of design requirements using design capability indices as metrics. *Engineering Optimization, 31*(5), 615–619.
25. Nellippallil, A. B., Mohan, P., Allen, J. K., & Mistree, F. (2020). An inverse, decision-based design method for robust concept exploration. *Journal of Mechanical Design, 142*(8), 081703.
26. Mistree, F., Bras, B., Smith, W. F., & Allen, J. K. (1996). Modeling design processes: A conceptual, decision-based approach. *International Journal of Engineering Design and Automation, 1*(4), 209–221.
27. Miller, J. G. (1973). *Living systems.* McGraw-Hill.
28. Pahl, G., Beitz, W., Feldhusen, G., & Grote, K. H. (2014). *Engineering design: A systematic approach.* 3rd Edition, Springer Science & Business Media.
29. Bras, B. (1992). *Foundation for designing decision-based design processes.* Ph.D. dissertation, Univesity of Houston.
30. Regli, W. C., & Cicirello, V. A. (2000). Managing digital libraries for computer-aided design. *Computer-Aided Design, 32*(2), 119–132.
31. Ming, Z., Yan, Y., Wang, G., Panchal, J. H., Goh, C. H., Allen, J. K., & Mistree, F. (2016). Ontology-based executable design decision template representation and reuse. *Artificial Intelligence for Engineering Design, Analysis & Manufacturing, 30*, 390–405.
32. Panchal, J. H., Fernández, M. G., Paredis, C. J. J., & Mistree, F. (2004). Reusable design processes via modular, executable, decision-centric templates. AIAA/ISSMO multidisciplinary analysis and optimization conference. Albany, NY. Paper Number AIAA-2004–4601.
33. Gruber, T. R. (1993). A Translation approach to portable ontology specifications. *Knowledge Acquisition, 5*(2), 199–220.
34. Noy, N. F., Fergerson, R. W., & Musen, M. A. (2000). *The knowledge model of Protégé-2000: Combining interoperability and flexibility* In R. Dieng, & O. Corby (Eds.), Knowledge Engineering and Knowledge Management Methods, Models, and Tools. EKAW 2000. Lecture Notes in Computer Science (Vol 1937). Springer.
35. Tudorache, T. (2006). *Employing Ontologies for an Improved Development Process in Collaborative Engineering*. Ph.D. dissertation, Technical University of Berlin.
36. Peraketh, B., Menzel, C. P., Mayer, R. J., Fillion, F., & Futrell, M. T. (1994). Ontology capture method (IDEF5). Knowledge Based Systems Inc.
37. Noy, N. F., & McGuinness, D. L. (2001). *Ontology development 101: A guide to creating your first ontology.* https://www.researchgate.net/publication/243772462_Ontology_Development_101_A_Guide_to_Creating_Your_First_Ontology.

38. Uschold, M., & Gruninger, M. (1996). Ontologies: Principles, methods and applications. *The Knowledge Engineering Review, 11*(2), 93–136.
39. Shah, J. J., & Mantyla, M. (1995). *Parametric and feature-based CAD/CAM: Concepts, techniques an applications.* Wiley.
40. Coyne, R. D. D., Rosenman, M. A., Radford, A. D., Balachandran, M., & Gero, J. S. (1990). *Knowledge-based design systems.* Addison-Wesley Pub. Co.
41. Finger, S., & Dixon, J. R. (1989). A review of research in mechanical engineering design. Part II: Representations, analysis, and design for the life cycle. *Research in Engineering Design, 1*(2), 121–137.
42. Verhagen, W. J. C., Bermell-Garcia, P., Van Dijk, R. E. C., & Curran, R. (2012). A critical review of knowledge-based engineering: An identification of research challenges. *Advanced Engineering Informatics, 26*(1), 5–15.
43. Rocca, G. L. (2012). Knowledge based engineering: Between AI and CAD. Review of a language based technology to support engineering design. *Advanced Engineering Informatics, 26*(2), 159–179.
44. Jakiela, M. J., & Papalambros, P. Y. (1989). Design and implementation of a prototype 'Intelligent' CAD system. *Journal of Mechanisms Transmission & Automation in Design, 111*(2), 252–258.
45. Sapuan, S. M. (2001). A knowledge-based system for materials selection in mechanical engineering design. *Materials & Design, 22*(8), 687–695.
46. Seepersad, C. C., Pedersen, K., Emblemsvåg, J., Bailey, R. R., Allen, J. K., & Mistree, F. (2006). The validation square: How does one validate design methods? In W. Chen, K. Lewis, & L. Schmidt (Eds.), *Decision-based design: Making effective decisions in product and systems design,* Chapter 25 (pp. 305–326). ASME Press.

Chapter 3
Ontology for Decision Support Problem Templates

We suggest that there are two primary decisions, namely, selection and compromise. These decisions can be mathematically modeled and coupled to model a process for designing a complex system. In this chapter, we deal with modeling and capturing knowledge related to selection, compromise, and coupled hierarchical decisions in design using template-based ontological methods. The knowledge models form the core of the knowledge base of the decision support platform. Three small examples, namely, a light switch cover plate assembly, a pressure vessel, and a portal frame are used to test the efficacy of the knowledge models. A summary of this chapter is presented in Table 3.1. The mapping of the sections to the components (topics) discussed in this chapter is presented in Row 2 of Table 3.1.

3.1 Frame of Reference

As indicated in Chap. 2, ontologies are formal specifications of terms and relations [1] and are widely utilized in the engineering domain for knowledge modeling, which includes (1) ***knowledge sharing***, (2) ***knowledge retrieval,*** and (3) ***knowledge documentation*** **for reuse**. The usefulness of an ontology to represent the common understanding of a domain and to share and exchange information among applications is embodied in knowledge sharing. Lu et al. [2] utilize ontology to capture geometric constraint specifications facilitating data exchange among various product development systems. Accordingly, Barbu et al. [3] create "OntoSTEP" to facilitate manufacturing utilizing an ontology that transforms digital models with geometric information into semantically rich models involving behavior and function. The usefulness of an ontology in representing complex relations among concepts and in constructing a reference network for knowledge retrieval is embodied in knowledge retrieval. For example, Li et al. [4] use ontology as an intelligent indexing scheme to create unstructured document information repositories and retrieve high recall and precision information. Liu et al. [5] model product families utilizing ontology and

© The Author(s), under exclusive license to Springer Nature Switzerland AG 2022

Z. Ming et al., *Architecting A Knowledge-Based Platform for Design Engineering 4.0*,
https://doi.org/10.1007/978-3-030-90521-7_3

Table 3.1 Summary interpretation of the problem investigated in this chapter

Elements	**What?**
Components	Modeling individual decisions and knowledge management, including the ontology for selection decisions (Sect. 3.2), the ontology for compromise decisions (Sect. 3.3), and the ontology for coupled decisions (Sect. 3.4)
Connectors	DSP technique, reusable and executable templates
Form	**How?**
Component roles	Provides the constructs for modeling individual decisions and capturing the related knowledge
Properties	Design
Relationship	Supporting selection, compromise, and coupled decisions
Rationale	**Why?**
Motivation	Provide knowledge support to assist designers in individual decision activities
Assumptions	There are three key types of decisions: selection, compromise, and combinations thereof
Interpretation	Understanding how selection, compromise, and coupled decisions are modeled, and how the associated knowledge is represented in a reusable and executable manner

recommend a setup for faceted information retrieval. The usefulness of the ontology in populating instances and documents specific domain knowledge is embodied in knowledge capture for reuse. For example, Witherell et al. [6] create an ontology to archive and reuse optimization knowledge involving assumptions, methods, results, etc. Rockwell et al. [7] create an ontology for archiving the rationale of design decisions.

In the context of PDSIDES, ontology is used to represent and document the knowledge related to selection, compromise, and coupled hierarchical decisions, thus making it possible for a designer to create new decision templates and reuse and execute existing templates in PDSIDES.

3.2 Ontology-Based Representation of the sDSP Template

In this section, we describe an ontology for representing the sDSP, the selection DSP template. We initially identify the requirements to model the knowledge related to selection decisions. Based on the requirements, we then discuss the structure of an sDSP template as the information model for select decisions. According to the structure of the sDSP template, an ontology is developed using Protégé, and its efficiency is demonstrated using a light switch cover plate material selection example.

3.2.1 *Requirements for Knowledge Modeling to Support Selection Decisions*

To facilitate designers formulating and analyzing selection decisions, we identify the following requirements for modeling the related knowledge:

- **Reusability**—the pursuit of design efficiency requires a model to be reused. In practice, many products or system designs fall into the category of variant or adaptive designs, where a large part of the underlying design model information can be reused without requiring the recreation of the model from scratch. For example, in an adaptive design scenario, only one part is changed, while the others remain the same, the actual design knowledge should be easily reused (with minor modifications) to support the design. The knowledge related to selection decisions, e.g., alternatives, attributes, etc., must be reusable to minimize the efforts and risk in the variant or adaptive design.
- **Executability**—the knowledge model of the decision selection needs to be executable on a computer. Encoding the linguistic and/or mathematical formulation of decision selection as a computational model is a feasible way to obtain executability. The disadvantage is that the codes are incomprehensible to most designers who are not programmers. Hence, there is a need for a selection knowledge model to be executable and simultaneously easy to communicate among designers. Therefore, the model must capture the semantics in the decision selection process.
- **Consistency—**the mathematical rigor of the selection decision formulation requires the computational model to be consistent with its definition. For example, the sum of the value of the scaling constants in a utility function should equal one. This is one of the rules for the mathematical model to maintain consistency. If the rules are violated during model reconfiguration, it will result in model inconsistency. Therefore, the computational model should support consistency checking and maintain that consistency.

3.2.2 *Information Model of Selection Decisions—The sDSP Template*

To address the requirements listed in Sect. 3.2.1, we propose the (utility-based) sDSP template, which is the information model of decisions selection (see Fig. 3.1). The declarative and procedural knowledge (corresponding to the "chips" and "wires" of Fig. 3.1, respectively) of the sDSP template is discussed in the remainder of this section. Declarative knowledge includes *alternative, attribute, utility function, evaluation, ranking,* and *post-solution analysis.* Their meaning in the context of a decision selection process is given as follows:

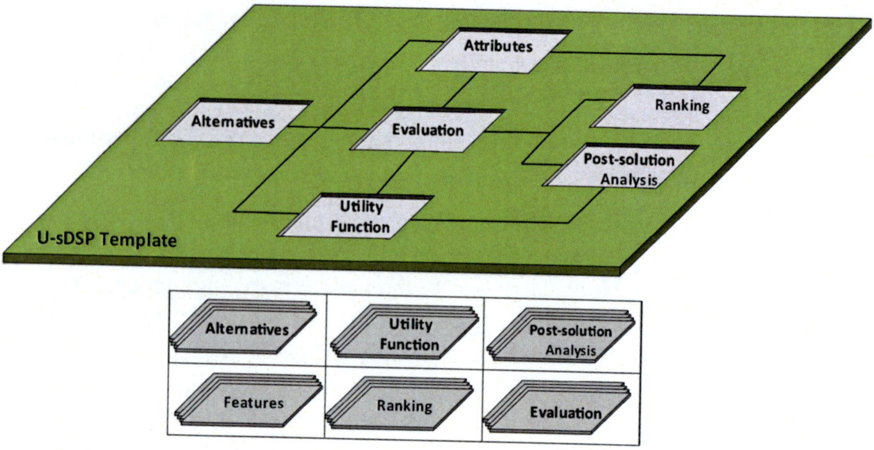

Fig. 3.1 The utility-based sDSP template [8]

(1) **Declarative Knowledge of the sDSP template**

- ***Alternatives***. A set of feasible options is considered to fulfill specific requirements.
- ***Attributes***. A set of performance measures for the alternatives that are used in the selection decision.
- ***Utility Function***. A function that takes the attributes of an alternative as the input and outputs the merit of that alternative. It indicates the decision-maker's preference toward a specific alternative.
- ***Evaluation***. The evaluation of the utility of every alternative.
- ***Ranking***. The rank of each alternative according to its utility.
- ***Post-solution Analysis***. The analysis of rank variation (or rank robustness) of the alternatives is due to the variation of the parameters in the utility evaluation process.

In addition to the declarative knowledge, we need to identify the procedural knowledge to enable the executability of the sDSP template. The procedural knowledge includes individual *utility function construction*, *expected utility calculation*, *multi-attribute utility function construction*, and *post-solution sensitivity analysis*. We discuss them as follows:

(2) **Procedural Knowledge of the sDSP template**

Individual utility function construction. The first step for evaluating the utility of the alternatives is to construct individual utility functions, which are the functions that accurately reflect the decision-maker's preference considering a particular attribute. The construction procedure consists of two steps. The first is to determine the attribute value levels that indicate specific utility values by answering lottery questions [9]. The second is to fit these identified attribute values into a function, which is deemed to be the decision-maker's utility function for that attribute. Typically, curve fitting is used as an automated method for transforming the answers to lottery questions to

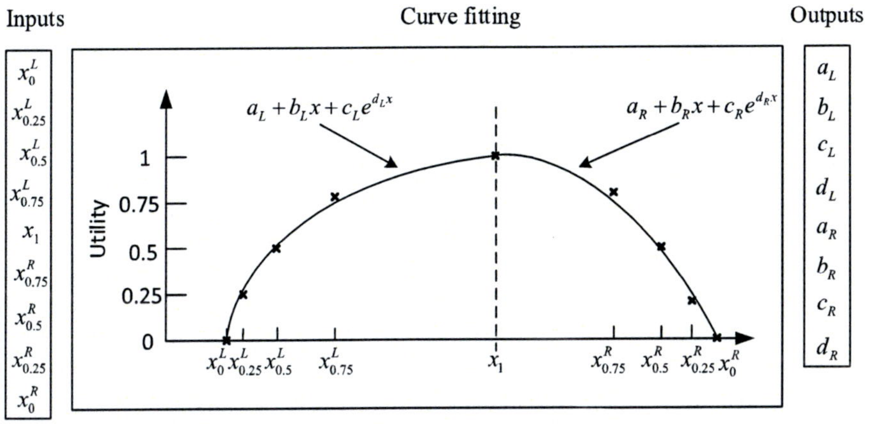

Fig. 3.2 Individual utility function construction [8]

individual utility functions. In Fig. 3.2, we show a specific input and output of the fitting method, with numbers x_0, $x_{0.25}$, $x_{0.5}$, $x_{0.75}$, and x_1 indicating the five attribute value levels, where superscripts indicate whether they are the *right-hand side* or *left-hand side* of the target and the subscripts indicate the utility value. For the non-monotonic (target-based) attributes, both the *right-hand side* and *left-hand side* attribute values are required as inputs, while concerning the monotonic attributes, either *right-hand side* (for monotonically decreasing attributes) or *left-hand side* (for monotonically increasing attributes) are needed. Coefficients a_L, b_L, c_L, d_L of the *left-hand side* utility function and (or) a_R, b_R, c_R, d_R of the *right-hand side* utility function represent the corresponding outputs of the method. This work assumes that the utility function is in the form $u(x) = a + bx + ce^{dx}$, where a, b, and c are coefficients.

Multi-attribute utility function construction. This procedure is designed to identify the multi-attribute utility function, which reflects the decision-maker's requirements as a combination of multiple conflicting attributes, through the integration of individual utilities. The core concept is that the multi-attribute utility function scaling constants k_i $(i = 1, \ldots n)$ [9, 10] are calculated by solving a system of n linear equations (see Fig. 3.3). In Fig. 3.3, A_{i^0}, A_{i^*}, $A_{i^{**}}$, and $A_{i^?}$ are various attribute levels i, with A_{i^0} being the poorest level with zero utility, A_{i^*} and $A_{i^{**}}$ levels conform to $A_{i^{**}} \succ A_{i^*}$ (it is assumes that $A_{i^{**}}$ refers to utility value 0.55 and A_{i^*} to 0.45, thus $A_{i^{**}}$ is favored to A_{i^*}), $A_{i^?}$ level is determined by the decision-maker to establish equivalent attribute combinations so that $(A_{1^?}, A_{2^*}, \ldots, A_{n^*}) \sim (A_{1^0}, A_{2^{**}}, \ldots, A_{n^{**}})$ posing the same utility as the decision-maker is indifferent. If both the additive and the multi-utility independence properties [9] hold, the multi-utility function adapts to the additive form $U = \sum_{i=1}^{n} k_i u_i(A_i)$. Based on the additive utility function form and the equivalent attribute combinations, multiple linear equations can be set up to represent the corresponding same utilities. These equations create an equation system involving n-variables, from which the k-values are determined. In this system of equations, $n - 1$ equations are set using the equivalent attribution combinations

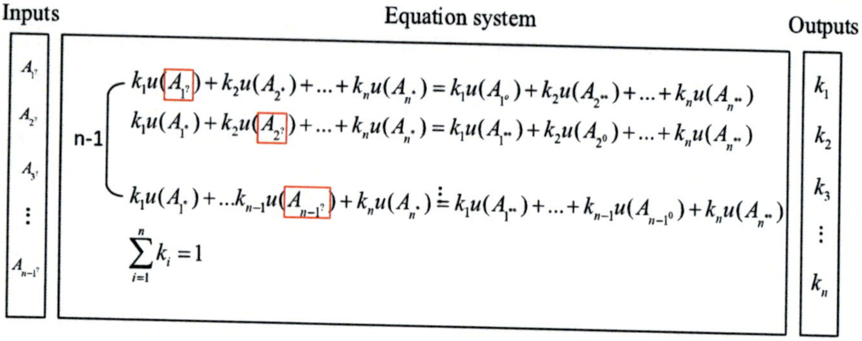

Fig. 3.3 Multi-attribute utility function construction [8]

determined by the decision-maker, and the remaining one is defined as $\sum_{i=1}^{n} k_i = 1$, indicating that the k-values sum to one. In the context of the sDSP template, solving this equation system resembles an automatic process transferring the specifications of some attribute levels into the multi-attribute function. The inputs of this equation system are the $A_{1?}, A_{2?}, \ldots, A_{n-1?}$ levels of the $n-1$ attributes, while the outputs $k_1, k_2, \ldots, k_n$ are the scaling constants.

Expected utility calculation. This process is to evaluate the expected utility of each alternative and identify the ranking of preferences. For a particular alternative, the expected utility is challenged first by estimating its expected utility by considering each attribute employing individual utility functions. Then, the overall expected utility is calculated by employing the multi-attribute utility function (Fig. 3.4). For an individual utility calculation, the expected utility of an alternative with respect to a target-based non-monotonic attribute is used as an example in the figure. For this example, the alternative's attribute range is assumed to be bounded between the attribute's *left-hand side* and *right-hand side* acceptable values, namely, x_0^L and x_0^R,

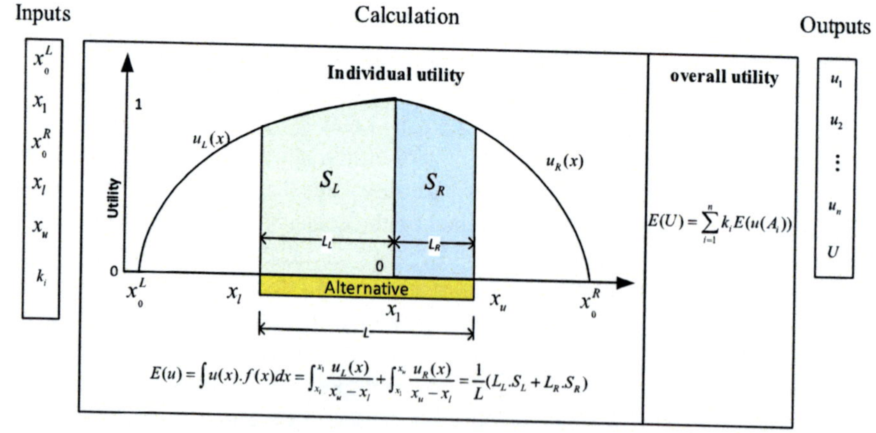

Fig. 3.4 Expected utility calculation [8]

respectively. Additionally, the distribution is assumed to be uniform with a probability density of $f(x) = \frac{1}{x_u - x_l}$, where x_u and x_l are the alternative's attribute upper and lower bounds, respectively. Relying on equation $E(u) = \int u(x) \cdot f(x)dx$, the individual expected utility is $E(u) = \frac{1}{L}(L_L \cdot S_L + L_R \cdot S_R)$, with L being the alternative's attribute range length, L_L and L_R the length measured from x_l to x_1 and x_1 to x_u, respectively. S_L and S_R denote the *left-hand side* and *right-hand side* areas bounded by the x-axis, which are the attribute value and utility function curve bounds, respectively. This equation is critical to generate variations in the post-solution analysis which is introduced later.

The results of the individual expected utility estimations are input to $E(U) = \sum_{i=1}^{n} k_i \cdot E(u_i(A_i))$ to generate the overall expected utility of the alternative. In the context of the sDSP template, the calculation process is automatic, with the inputs involving (i) x_0^L, x_0^R and x_1, with the first two being the acceptable attribute levels, and the last the ideal attribute level, (ii) x_l and x_u denote the alternative's attribute value bounds, and (iii) the k-values. Accordingly, the outputs involve the individual expected utility values $u_1, u_2, \ldots, u_n$, and the overall expected utility U.

Post-solution sensitivity analysis. This process is to test the sensitivity of the results to parameter variations and enhance the decision-maker's confidence in selecting the most promising alternative. Firstly, the top-two alternatives (or even more if the expected utilities are close to each other) are selected to test whether the parameter variations influence the ranking of the two alternatives. The expected utility estimation process of Fig. 3.4 highlights that a parameter variation is only possible in the $E(U)$ equation, which is used to calculate the overall expected utility, and equation $E(u)$ is used to calculate the individual expected utility. In Fig. 3.5, Δk is the parameter variation of $E(U)$, and Δx and ΔL denote the parameter variation of $E(u)$. Δk represents scaling variation, influencing the overall expected utilities of

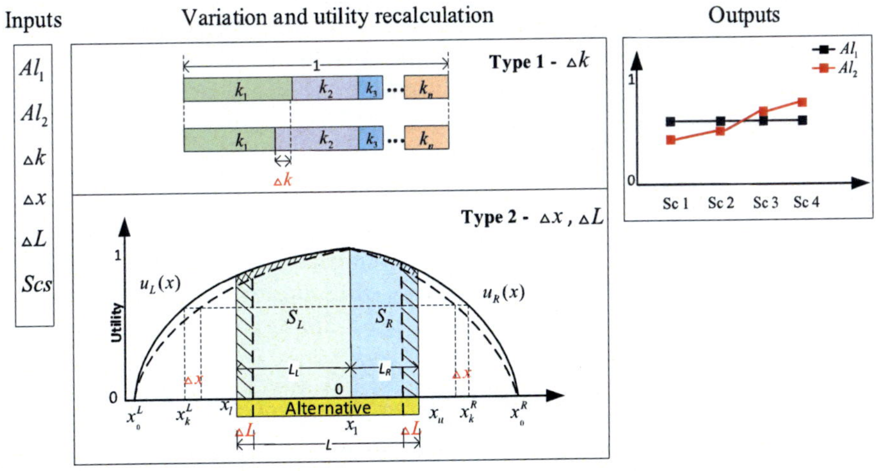

Fig. 3.5 Post-solution sensitivity analysis [8]

all the alternatives. Δx represents the variation of the individual utility function influencing the individual expected utilities of all the alternatives. If Δx moves toward the target value x_1, the alternative, is more risk-prone, while moving in the opposite direction indicates it is more risk-averse. Finally, ΔL represents the attribute value range variation of a specific alternative (increasing or reducing), and it only affects the individual expected utilities of that alternative. For the sensitivity analysis, the decision-maker may formulate various scenarios utilizing a constant decrement or increment for Δk, Δx, and ΔL, e.g., 5%, and then recalculates $E(U)$ and $E(u)$ to obtain the corresponding expected utility values of the top-two alternatives. Then, the ranking changes are displayed using visualization tools, i.e., the line chart illustrated in Fig. 3.5. In the context of the sDSP template, post-solution analysis is identified as an automatic procedure that transfers the decision-maker's specification of parameter variations to the sensitivity in the ranking of alternatives. The inputs require: (i) the top-two alternatives Al_1, Al_2, (ii) the parameter variation, i.e., Δk, Δx, and ΔL, and (iii) the number of scenarios (Scs). The output of this process is the corresponding sensitivity graphs.

3.2.3 sDSP Template Ontology Development

In this section, we deal with knowledge representation, i.e., developing an ontology relying on the sDSP template. The Slots, Classes, and consistency rules establishing a frame-based ontology are:

(1) **Class Definition**

The six "chips" shown in Fig. 3.1 are the key concepts that constitute the main structure of the u-sDSP template ontology and are explicitly defined as Classes in this section. In addition to them, seven additional Classes referred as *U-SDSPTemplate, MUtilityFunction, IUtilityFunction, IAttributeAssessment, MAssessment, UtilityCalculation,* and *Coefficient* are defined to capture the knowledge that adds to the semantic richness and integrity of the ontology. The definitions of the Classes are shown in Table 3.2.

(2) **Slot Definition**

The semantic relationships between Classes are captured using Slots in a frame-based ontology. Based on the utility-based sDSP construct and the utility-based sDSP template structure, the data Slots and object Slots of the ontology are defined as shown in Tables 3.3 and 3.4, respectively.

(3) **Consistency Rule Definition**

By restricting the populated Instances based on their definition, an ontology must preserve its consistency. Ontology inconsistency commonly occurs during the instantiation process and involves populating the ontology employing detailed data in the modification stage, where the original Instances are modified. Hence, it is critical

Table 3.2 Utility-based sDSP template ontology classes [8]

Class	Definition
U-SDSPTemplate	A formulation of a selection decision problem with multiple conflicting attributes under uncertainty. Integration of all the template modules and the associated information
Alternative	A feasible option considered to fulfill certain requirements
Attribute	A quality or feature regarded as a characteristic part of an alternative
UtilityFunction	A function used to indicate the decision-maker's preference with respect to attribute(s)
MUtilityFunction	A subclass of UtilityFunction, whose function values indicate the decision-maker's preference with respect to the combination of these attributes
IutilityFunction	A subclass of UtilityFunction, whose function values indicate the decision-maker's preference with respect to that single attribute
Evaluation	The making of judgments about the utility values of the alternatives. Including the information captured by Classes *IattributeAssessment, Massessment, and UtilityCalculation*
IAttributeAssessment	The assessment of the decision-maker's preference with respect to a single attribute. Capturing the information of the inputs shown in Fig. 3.4
MAttributeAssessment	The assessment of the decision-maker's preference with respect to multiple attributes. Capturing the information of the inputs shown in Fig. 3.5
Coefficient	The coefficient in the utility function of the form $u(x) = a + bx + ce^{dx}$
UtilityCalculation	The calculation of the expected utility of each alternative with respect to each attribute. This is used to capture the information of the inputs shown in Fig. 3.6
Ranking	The order of the decision-maker's preferences with respect to the alternatives
PSAnalysis	The analysis of the sensitivity of the decision-maker's preferences with respect to the alternatives considering some type(s) of parameter variation

to detect inconsistency and inform designers to resolve it. A common technique checking for ontology consistency is rule-based reasoning, with the corresponding rule-set for the utility-based sDSP template ontology presented in Table 3.5. The rule engine JESS (Java Expert System Shell) [11] is based on the Java platform. To comply with JESS, the rules are appropriately defined, with Rule 2 being a typical example:

*(defrule MAIN::rule_2 (object (is-a : IAttributeAssessment) (:*attributeLevel 0*)*
*(:*side Left-hand-side*) (:value ?x) (:attribute ?a))*

=>

(if (neq ?x (slot-get ?a lowerBound)) then (printout t WARNING_2 crlf)))

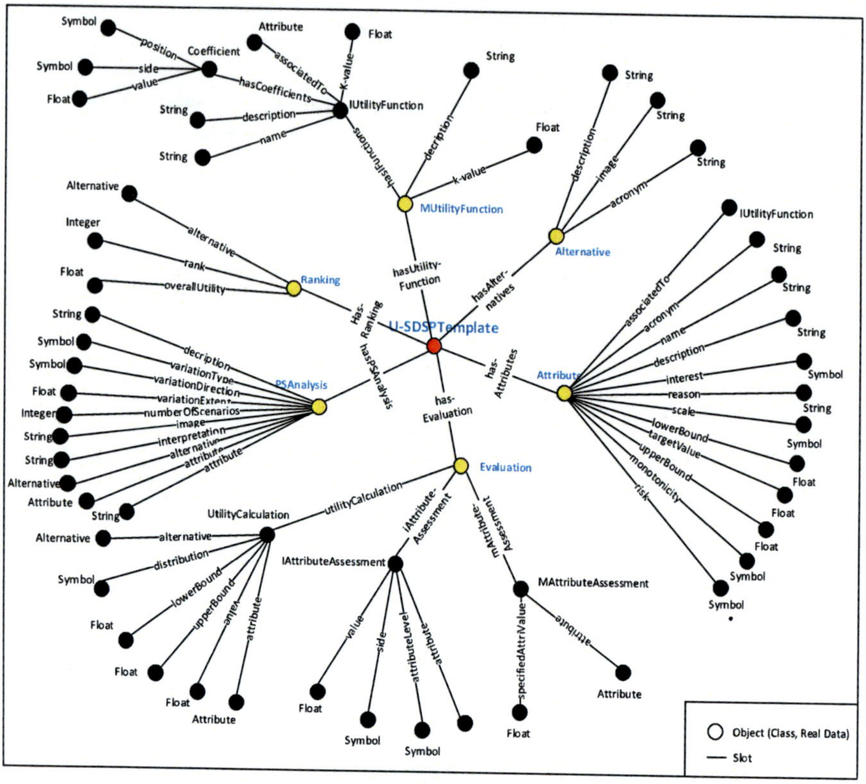

Fig. 3.6 Overview of the utility-based sDSP template ontology [8]

Hence, if any "IAttributeAssessment" Instance violates Rule 2, the reasoner sends the designer working on that utility-based sDSP template the message "WARNING_2" about the inconsistency.

(4) Ontology Overview

In Fig. 3.6, we show the u-sDSP template ontology structure integrating the concepts and relations. Essentially, this is a network structure where nodes represent Classes and real data. Different Classes and data are linked by Slots (including object and data Slots), represented as lines with the Slot names placed at the middle. A red node indicates the ontology's core Class, representing the template's information entry. The six yellow nodes refer to ontology's key components, representing the template's "chips" (see Fig. 3.1). The ontology modeling process exploits the Protégé tool [12] developed by Stanford University to create and edit ontologies and populate Instances based on ontologies. Additionally, we use the JessTab plugin [13] to attach JESS to Protégé and check the ontology's consistency. We also develop a plugin to execute the populated instances by employing Java function calls.

Table 3.3 Utility-based sDSP template ontology data slots [8]

Slot name	Definition	Type
name	Name of an Instance	String
description	Description of an Instance	String
problemStatement	A statement of a selection problem	String
acronym	An acronym of an attribute or alternative	String
interest	Whether an attribute is of interest for the selection or not ("Yes" or "No")	Symbol
reason	The reason why an attribute is considered or not	String
risk	The decision-maker's attitude about risk toward an attribute ("Aversion", "Proneness" or "None")	Symbol
monotonicity	The monotonicity of the decision-maker's preference about an attribute ("Target", "Increasing", or "Decreasing")	Symbol
scale	The scale of an attribute ("Ratio" or "Interval")	Symbol
unit	The unit of an attribute's measurement	String
lowerBound	The lower bound of a general attribute or an alternative's attribute	Float
upperBound	The upper bound of a general attribute or an alternative's attribute	Float
targetValue	The ideal value of an attribute	Float
attributeLevel	A level of an attribute ("0", "0.25", "0.5", "0.75", or "1")	Symbol
side	The side of the target ("Left-hand-side" or "right-hand-side")	Symbol
value	The value of an attribute or utility	Float
specifiedAttriValue	An attribute value specified by the decision-maker in creating an equivalence	Float
k-value	The scaling constant associated to an individual utility function	Float
position	The position of an efficient in $u(x) = a + bx + ce^{dx}$ ("a", "b", "c", or "d")	Symbol
distribution	The distribution of the attribute value ("Uniform", "Normal" etc. Default is Uniform)	Symbol
rank	The rank of a particular alternative	Integer
overallUtility	The overall utility of an alternative with respect to all the attributes	Float
variationType	The type of variation in post-solution analysis. ("Δk", "Δx", or "ΔL")	Symbol
variationDirection	The direction of variation ("Increment" or "Decrement")	Symbol
variationExtent	The extent of variation (the unit is "%")	Float
image	The (path of) graph that captures the responding sensitivity	String
interpretation	The interpretation of a sensitivity graph	String

Table 3.4 Utility-based sDSP template ontology object slots [8]

Slot name	Definition	Type
hasAlternatives	Link a *U-SDSPTemplate* to a set of *Alternatives*	Alternative
hasAttributes	Link a *U-SDSPTemplate* to a set of *Attributes*	Attribute
hasUtilityFunction	Link a *U-SDSPTemplate* to an *UtilityFunction*	UtilityFunction
hasEvaluation	Link a *U-SDSPTemplate* to an *Evaluation*	Evaluation
hasRanking	Link a *U-SDSPTemplate* to a *Ranking*	Ranking
hasPSAnalysis	Link a *U-SDSPTemplate* to a *PSAnalysis*	PSAnalysis
associatedTo	Interrelate two different Classes	Instance
HasCoefficients	Link an *IutilityFunction* to a set of *Coefficients*	Coefficient
hasIFunctions	Link a *MutilityFunction* to a set of *IUtilityFunctions*	IUtilityFunction
iAttributeAssessment	Link an *Evaluation* to a set of *IAttributeAssessment*	IAttributeAssessment
mAttributeAssessment	Link an *Evaluation* to a set of *MAttributeAssessment*	MAttributeAssessment
utilityCalculation	Link an *Evaluation* to a set of *UtilityCalculations*	UtilityCalculation
attribute	Link a *MAttributeAssessment*, *IAttributeAssessment*, or *UtilityCalculation* to a (set of) *Attribute*(s)	Attribute
alternative	Link an *UtilityCalculation* or a *Ranking* to an *Alternative*	Alternative

Table 3.5 Utility-based sDSP template ontology consistency rules

Rule 1	Every "Attribute" Instance must conform to lowerBound ≤ targetValue ≤ upperBound
Rule 2	Every "IAttributeAssessment" Instance must conform to: (attributeLevel = 0 & side = Left-hand-side) => (value = attribute.lowerBound)
Rule 3	Every "IAttributeAssessment" Instance must conform to: (attributeLevel = 1) => (value = attribute.lowerBound)
Rule 4	Every "IAttributeAssessment" Instance must conform to: (attributeLevel = 0 & side = Right-hand-side) => (value = attribute.upperBound)
Rule 5	Every "IUtilityFunctions" Instance must conform to 0 ≤ k-value ≤ 1
Rule 6	Every "IUtilityFunctions" Instance must conform to: $\sum_0^n k_i = 1$
Rule 7	The number of Instances in the Slot "utilityCalculation" should be equal to alternatives × attributes
Rule 8	Every "UtilityCalculation" Instance must conform to: lowerBound < upperBound

3.2.4 *Test Example—Material Selection for a Light Switch Cover Plate*

In this section, a Rapid Prototyping (RP) resource selection problem is used as an example to illustrate the usefulness of the utility-based sDSP template ontology in terms of facilitating the reuse and execution of the embedded knowledge. The example is an extension of the problem considered by Fernández and coauthors [10] who use it to illustrate the process of formulating and solving a specific utility-based sDSP. First, an instance is created using the original information, then three knowledge-reuse scenarios are presented, and finally, we give a summary and discussion.

(1) **Populating an Original Selection Instance**

Rapid prototyping (RP) affords the opportunity to reduce operating costs and time to develop new, redesigned, and customized products. Selecting the proper resources, e.g., materials and manufacturing processes, is a vital attribute of RP technology. The example used to illustrate the method involves the selection of material-process combinations to make a light switch cover plate assembly (as shown in Fig. 3.7) using RP technology [10]. The primary prototype objectives, presented in decreasing importance order, are (1) validating the function product, specifically concerning the snap-fitting of components as highlighted in Fig. 3.7, (2) determining the fitting or tolerance closeness of the two interfacing components, (3) obtaining a feel for the product, and (4) visually and physically confirming the 3D interface integrity. To manage these objectives, we consider the four alternatives and five attributes

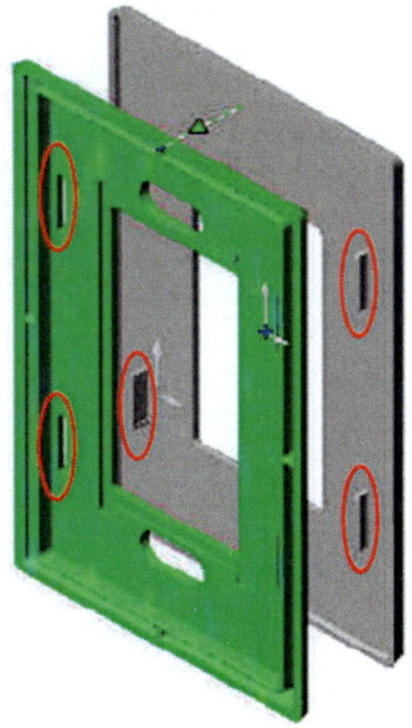

Fig. 3.7 Light switch cover plate assembly [10]

Table 3.6 Alternatives and attributes considered for selection [10]

Alternative		Attribute				
Process	Material	Ratio			Interval	
		Tensile strength (50-65-75) MPa	Young's modulus (1500-2137-2600) MPa	Flexural strength (70-95-120) MPa	Detail capability (−0.4–0.9) mm	Accuracy (−0.02–0.1)
SLA250	DSM7110	44–69	1758–2413	59–110	0.45–0.55	0.04–0.05
SLA3500	DSM8120	23–29	633–773	23–29	0.45–0.55	0.04–0.05
FDM1650	P400	31–37	2234–2730	58–72	0.45–0.55	0.12–0.14
MJM2100	TJ75	9–11	90–110	9–11	0.67–0.83	0.12–0.14
Distribution: Uniform						

presented in Table 3.6. The former involves four material-process combinations (for further details, the reader is referred to [10]). The attributes are a mixed setup involving ratio and interval scales, i.e., *Young's Modulus, Tensile Strength*, and *Flexural Strength are* measured by ratio scales, and Interval scales measure *Accuracy* and *Design Capability*. The designer's requirements toward these attributes are a mixture of monotonicity and non-monotonicity, e.g., *Accuracy* is an attribute posing a monotonically decreasing requirement characteristic, while *Tensile Strength* is a non-monotonic target-based requirement attribute. For example, "65 MPa" is the target for *Tensile Strength,* as presented in Table 3.6. The alternatives' attribute values are uncertain and bounded with a lower and an upper bound, e.g., "44–69 MPa" is the range value of the attribute *Tensile Strength* for the alternative *SLA250-DSM7110.* It is assumed that all attribute values are subjected to uniform distributions. From the data shown in Table 3.6, a designer selects the most promising alternative.

It is assumed that the designer has appropriately formulated the problem employing the utility-based sDSP construct. Based on the formulation, an RP resource selection Instance is populated in Protégé, as illustrated in Fig. 3.8. In Fig. 3.8, the left panel is the Class browser listing all Classes, the middle panel is the Instance browser presenting all Instances associated with a selected Class, and the right panel is the Instance editor to create and edit an Instance. For this example, Slots including both data and object Slots, e.g., "problemStatement", "hasAlternative", "hasAttributes", etc., of the RP resource selection Instance are created based on the ontology presented in Sect. 3.2.3 and the problem itself. Given the automated procedures identified in Sect. 3.2.2, some slot values, e.g., Slot "k-value", representing the outputs of the relevant procedures, e.g., multi-attribute utility function construction, are automatically generated utilizing the Java function calls based on the encoded logic. The selection process produces the ranking order (in increasing order) *SLA250-DSM7110, FDM1650-P400, SLA3500-DSM8120, MJM2100-TJ75.* The overall utility of the top alternative *SLA250-DSM7110* is 0.69, presented in the window of Fig. 3.8, and indicated with a red arrow.

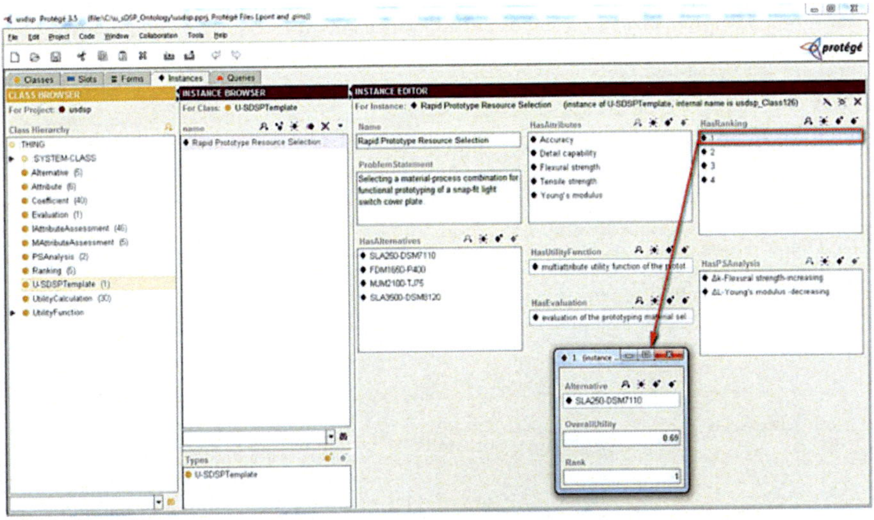

Fig. 3.8 Protégé ontology editor screenshot for the RP resource selection instance [8]

(2) Knowledge Reuse—Modification of the Original Instance

Scenario 1: New Alternatives Introduction. In this scenario, we consider a new combination of the *SLA3500* process provided by 3D Systems and the *SL7510* material by Vantico AG as an alternative to prototype the switch cover plate assembly. Specification of the new alternative *SLA3500-SL7510* is presented in Table 3.6. The problem transfers from a four-alternative problem to a five-alternative problem, where the designer has to reformulate the new problem and reselect. Concerning sDSP ontology, given that the knowledge related to the attributes, the multi-attribute utility function, the individual utility functions, and the remaining parameters are well documented in the original selection decision template Instance, the selection process can exploit this knowledge after modifying it appropriately. The inconsistency brought by the modification is managed using a rule-based consistency checking mechanism introduced in Sect. 3.2.3. For a detailed example, the reader is referred to [14]. Here, the key modification is adding a new alternative exploiting Table 3.7 and then updating the ranking, as presented in Fig. 3.9. Specific steps are as follows:

(1) Specify the general information—provide information in Slots “acronym”, “description”, and “image” to instantiate a new alternative.
(2) Specify attribute value ranges—provide information in Slots “lowerBound”, “upperBound”, and “distribution” concerning a particular attribute. In this example, we only present the requirement considering the attribute “Tensile Strength”, while other attribute value ranges can be similarly determined.
(3) Apply the new alternative in Slot “hasAlternatives”.
(4) Update ranking—once all the compulsory information is input, the ranking is automatically updated. The experimental results reveal that the new alternative

Table 3.7 The alternative SLA3500-SL7510 specification [8]

Alternative		Attribute				
Process	Material	Ratio			Interval	
		Tensile strength (50-65-75) MPa	Young's modulus (1500-2137-2600) MPa	Flexural strength (70-95-120) MPa	Detail capability (−0.4–0.9) mm	Accuracy (−0.02–0.1)
SLA3500	SL7510	42.3–55.46	1877–2869	78–96	0.45–0.55	0.04–0.05
Distribution: Uniform						

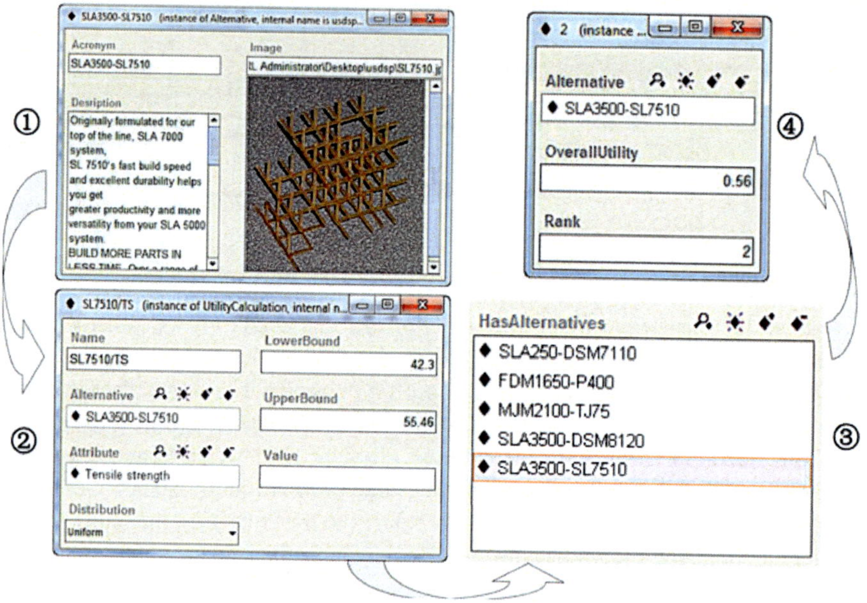

Fig. 3.9 Modifying the original Instance when new alternatives are introduced [8]

manages an overall utility of 0.56 and ranks second. Thus, it is not the best alternative.

Scenario 2: Introducing New Attributes. In this scenario, we consider a new attribute, i.e., *Flexural Modulus*, which measures a material's flexural stiffness modulus to evaluate the five alternatives. This attribute is necessary as it involves validating the stiffness requirements of the snap-fit design. Specifying this attribute involves the four facets presented in Table 3.8, (i) general information, i.e., scales, units, lower and upper unacceptable values, and target (ideal value), (ii) assessing the right-hand and left-hand side utility that is required to construct the attribute's utility function, (iii) the attribute levels to create equivalences to re-determine the attribute's "k-values", (iv) the *Flexural Modulus* range value per alternative that is employed to evaluate the expected utility concerning this attribute. Utilizing the specifications

Table 3.8 Specification of the attribute flexural modulus [8]

I	**Units**	**Scale**	**Lower unacceptable**		**Ideal**		**Upper unacceptable**		
	MPa	Ratio	1800		2344		2800		
II	**Left-Hand Side Utility (MPa)**				**Left-Hand Side Utility (MPa)**				
	0	**0.25**	**0.5**	**0.75**	**1**	**0.75**	**0.5**	**0.25**	**0**
	1800	1861.34	1995.45	2111.50	2344	2505.46	2623.34	2723.31	2800
III	**TS**		**YM**		**FS**	**FM**		**DC**	
	62 MPa		2030 MPa		90 MPa	2227 MPa		0.42 mm	
IV	**DSM7110**		**SL7510**		**DSM8120**	**P400**		**TJ75**	
	1710–2668 MPa		2374–2902 MPa		621–759 MPa	2358–2882 MPa		90–110 MPa	

introduced in Table 3.8, a designer needs to formulate a new six-attribute problem and reselect. As mentioned earlier, the majority of the documented knowledge can be reused after being adequately modified. Concerning sDSP ontology, the key modification is instantiating a new attribute and then updating the ranking, as shown in Fig. 3.10. Specific steps are as follows (the input of the steps come from Table 3.8):

(1) Specify the general information—provide information for Slots "name", "acronym", "description," etc., to instantiate a new alternative.
(2) Specify information for the individual attribute utility assessment—specify attribute values for the nine levels of utility. Window ③ in Fig. 3.10 has the specification of level 0.25.
(3) Specify information for the k-value assessment—specify attribute values for creating equivalences and determining the k-values of the multi-attribute utility function. Window ③ in Fig. 3.12 is the specification for the attribute *Flexural Modulus.*
(4) Specify the attribute ranges for each alternative—the attribute's upper and lower bounds per alternative are determined. The range of the alternative SLA3500-SL7510 is presented in window ④ of Fig. 3.10.

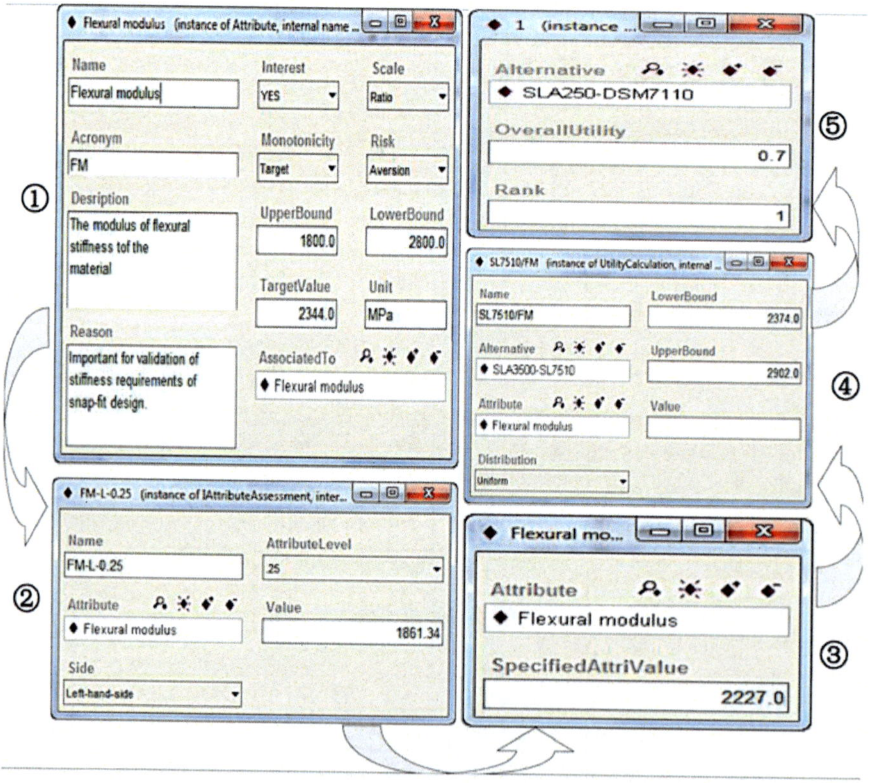

Fig. 3.10 Adjusting the original Instance when new attributes are introduced [8]

(5) Input the new attribute and update the ranking—the newly specified attribute *Flexural Modulus* is input in the "hasAttribute" Slot to update the ranking automatically. The results indicate that the top-two alternatives are *SLA250-DSM7110* and *SLA3500-SL7510*, present an overall utility of 0.7 and 0.56, respectively.

Scenario 3: Parameter Variation. In this scenario, some potential parameter variation needs to be considered to test the robustness (or sensitivity) of the selection and strengthen the designer' confidence in selecting the most promising alternative *SLA250-DSM7110* to prototype the assembly. The alternative ranked second (*SLA3500-SL7510,* 0.56) and is close to the top alternative (0.7). In this scenario, we vary the parameters and investigate whether the ranking is affected. Since post-solution sensitivity analysis is modeled as a u-sDSP template module, it can be performed independently by invoking the necessary information of other modules. In the ontological context, what needs to be done is simply instantiating an Instance of the Class "PSAnalysis" then plugging in the Slot "hasPSAnalysis", information in other Slots are unchanged. The instantiation of "PSAnalysis" includes specifying the Slots "alternative", "attribute", "variationType", "variationDirection", "variationExtent", "numberOfScenarios", etc., as defined in Table 3.3. The Instance Slot "image" captures the automatically generated graph based on the preceding specification, which is then utilized to visualize the change of ranking due to the variation. The designer interprets the result based on the image, then populates the Slot "interpretation", and finally documents the Instance. In Sect. 3.2.2, we identify three types of parameter variation, i.e., Δk*,* Δx*, and* ΔL*,* here we take ΔL, namely, the variation of attribute value range as an example. In the example, the original range (1877–2869) of the attribute *Young's Modulus* for alternative *SLA3500-SL7510* is gradually reduced (i.e., the uncertainty of this attribute is reduced) to test the responses of the alternative's overall utility and the relative ranking compared to the top one, as shown in Fig. 3.11. It can be seen in Fig. 3.11 that seven scenarios are scheduled for the testing and the attribute range is reduced by 5% in each scenario. The automatically generated image indicates that when the attribute range is reduced to 80% (represented by the fourth scenario in the image), the overall utility of the alternative *SLA3500-SL7510* increases from 0.56 to 0.6. Then it stabilizes at 0.6 despite a further range reduction. During the entire process, the alternative *SLA3500-SL7510* consistently ranks second. Thus, it is evident that the alternative's *SLA3500-SL7510* ranking is insensitive to reducing *Young's Modulus* uncertainty, and it is safe to select the most promising alternative, i.e., *SLA250-DSM7110*, even if the designer is confident about the *Young's Modulus* of *SLA3500-SL7510.*

3.3 Ontology-Based Representation of the cDSP Template

In this section, we develop an ontology for representing the cDSP template. First, we identify the requirements for modeling the knowledge related to compromise

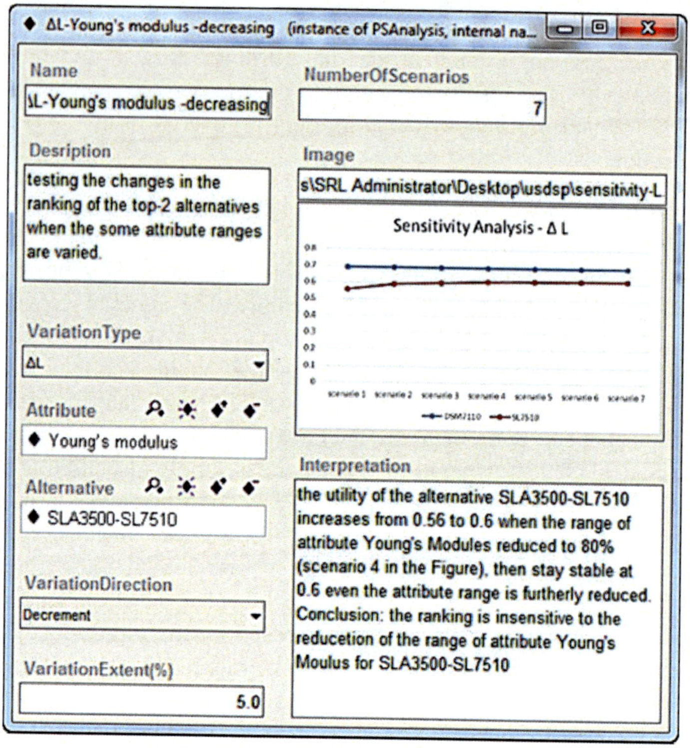

Fig. 3.11 Instantiation of the post-solution sensitivity analysis [8]

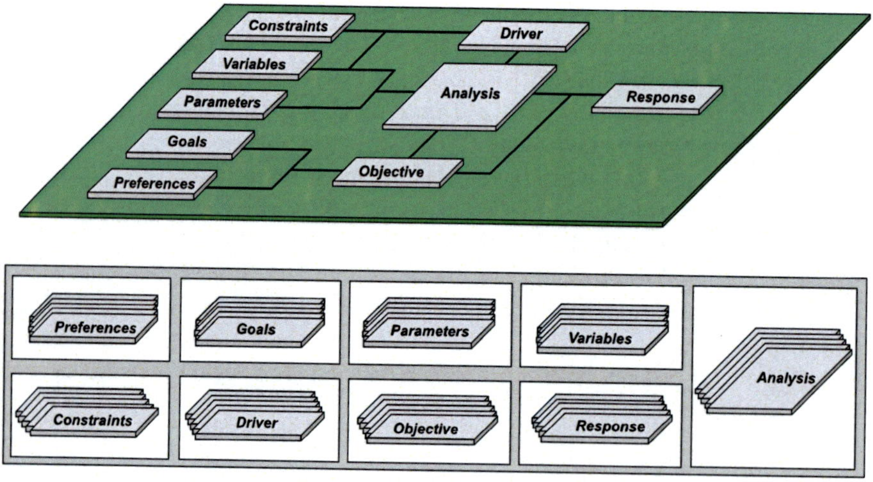

Fig. 3.12 The cDSP template [15, 16]

decisions. Second, we discuss the structure of a cDSP template for addressing the requirements. Third, we develop an ontology according to the structure of the cDSP template. Finally, we use as an example a pressure vessel design to test the ontology's utility.

3.3.1 Requirements for Knowledge Modeling to Support Compromise Decisions

According to the cDSP construct, many elements are involved in compromise decisions, such as constraints, goals, variables, parameters, etc. To facilitate designers formulating and analyzing compromise decisions, we identify the following requirements for modeling the related knowledge. These requirements are similar to what is identified for selection decisions in Sect. 3.2.1, but the context is different.

- **Reusability**—In the context of a compromise decision, reusability means that both domain-dependent knowledge (or declarative knowledge, e.g., variables, parameters, constraints, goals, weights, etc.) and domain-independent knowledge (or procedure knowledge, e.g., the algorithm for finding a good enough solution for the decision) should be reusable so that designers can quickly reuse existing knowledge (with some modification) in variant and adaptive designs, saving much time compared to re-implementing everything from scratch.
- **Executability**—In the context of a compromise decision, executability means that different decision elements are represented in a computer-interpretable manner to be integrated as a whole and executed to generate solutions. This is significant because solving compromise decisions is a computation-intensive process.
- **Consistency**—In the context of a compromise decision, consistency means that the elements as individuals and as a whole must keep being consistent with their definition in a mathematical rigorous cDSP construct. For instance, a cDSP variable should have an initial value lying between a lower bound and an upper bound, and the sum of the goal weights should equal one, etc.

3.3.2 Information Model of Compromise Decisions—The cDSP Template

To address the requirements for knowledge modeling and support compromise decisions, we model compromise decisions using the cDSP template, as shown in Fig. 3.12. Declarative knowledge in the model includes *Constraints*, *Variables*, *Parameters*, *Goals*, *Preferences*, *Driver*, *Analysis*, *Objective*, and *Response*.

- ***Constraints***. A hard rules (relationships) collection that must be satisfied to ensure the design feasibility.

- ***Goals***. A soft rules (relationships) collection that can be violated to incorporate the designer's wishes on the system.
- ***Variables***. A variable quantities collection that defines the system under consideration.
- ***Parameters***. A fixed quantities collection that is given before the decision and preserved during the decision process.
- ***Preferences***. The weights (or levels) assigned to the goals by the designer reflecting the relative importance of the goals.
- ***Objective***. The overall objective of the decision is to consider the preferences of the goals.
- ***Analysis***. The analysis of deviations of the solution points in the feasible design space.
- ***Driver***. The interface to the problem-solvers running the analysis codes.
- ***Response***. The actual response to a given cDSP template specification.

As discussed in Chap. 2, the adaptive linear programming (ALP) [17] algorithm and a computational tool called DSIDES are used to solve cDSP. Therefore, in the context of the cDSP template, the procedural knowledge is embedded in the DSIDES program. In several adaptive and variant design scenarios, the majority of the previous decision information can be reused to make decisions on the current design due to the similarity of the underlying design concepts. In the context of the cDSP template, this refers to reusing the majority of the modules in future designs. However, to successfully reuse the modules, partial adaptation is required to meet the design changes. Our goal in this section is to determine the cDSP model modification types that can be made in future design scenarios. The cDSP construct provides designers high flexibility to adjust their decision model to meet the requirements of the explicit problem investigated. The major modification types are: (i) altering the design variables' cardinality, (ii) altering the parameter values, and (iii) altering the goals and constraints' cardinality. A detailed description of these modification types is presented next in detail, along with the effects each modification type implies on the design space (illustrated in Fig. 3.13).

- **Type I: Altering the design variables cardinality**—this modification type involves adding new variables and removing current variables. The first case is implemented considering existing fixed value parameters as variables. For instance, in Fig. 3.13, x_3 while previously considered as a parameter, it is now reformulated as a variable, transforming the design space from a two-dimensional to a three-dimensional space. As shown in Fig. 3.13, the latter is realized by assigning a fixed value to an existing variable, and thus the value of the previous variable x_3 is fixed, and the design space transforms from three-dimensional to two-dimensional.
- **Type II: Altering the parameter's value**—this involves: (i) Altering the value assigned to the goal's **weights** (e.g., w_i's value in the deviation function is modified in Fig. 3.13) influencing the goals' achievements (ii) Altering the values assigned to the equations' **coefficients** (e.g., the coefficient value a of goal G_1 and coefficient value D of constraint C_3 are altered in Fig. 3.13), resulting in changing

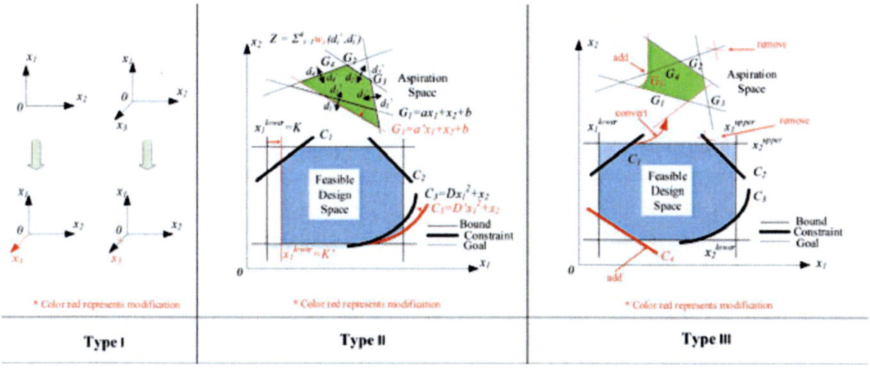

Fig. 3.13 Three types of the cDSP model modification [14]

the curve shapes affecting the aspiration and feasible design space, (iii) Altering the values assigned to the variables' **bounds** (e.g., the x_1^{lower} value is altered in Fig. 3.13), resulting in shrinking or expanding the feasible design space.

- **Type III: Modifying the goals' and constraints' cardinality**—this includes: (i) adding new goals, e.g., G_5 in Fig. 3.13, (ii) removing existing goals, e.g., G_4 in Fig. 3.13, (iii) adding new constraints, e.g., C_4 in Fig. 3.13, (iv) removing existing constraints, e.g., C_2 in Fig. 3.13, (v) converting existing goals to constraints or vice versa, e.g., in Fig. 3.13 constraint C_1 is converted to a goal. Adding new goals and converting existing constraints to goals shrinks the aspiration space, while removing existing goals and converting existing goals to constraints expands the aspiration space. Accordingly, adding new constraints and converting existing goals to constraints shrinks the feasible design space, and removing existing constraints and converting existing constraints to goals expands the feasible design space.

In practice, several modification types are simultaneously needed based on the requirements, e.g., in an adaptive design scenario, designers may need to change a previously fixed parameter to a variable, alter the variables' bounds, add new constraint to the problem, and hence all three cDSP model modification types may be needed simultaneously. In our cDSP ontology presented in Sect. 3.3.3, the modification is facilitated by editing the template instances in the data and object slots.

3.3.3 Ontology Development for the cDSP Template

According to the cDSP template that we present in Sect. 3.3.2, we develop an ontology for representing the associated knowledge in this section. The development process includes: (i) identifying the key concepts and formally expressing them as classes, (ii) identifying the relationship between the concepts and formally expressing them as slots, (iii) identifying and formally expressing consistency rules. Finally, presenting the complete cDSP template ontology structure.

(1) **Identifying the Concept**

A vocabulary of terms commonly is needed to describe the concepts in a domain. In Sect. 3.3.3, we identify the required terms to model the cDSP template, including *Goal, Constraint, Variable, Parameter, Analysis, Preference, Response,* and *Driver.* These terms are exploited and reused by expressing them as classes of an ontology. Six additional classes referred to as *Problem, Behavior, cDSPTemplate, Function, History, and Quantity,* are introduced to capture the information that adds to the ontology integrity and semantic richness. The relevant class expressions are presented in Table 3.9.

(2) **Relation Definition**

In an ontology, the semantic relations between concepts are captured by exploiting slots. Utilizing the mathematical construct of the cDSP, the cDSP's ontology data and object slots are defined in Tables 3.10 and 3.11, respectively (Table 3.10).

(3) **Maintaining Consistency**

Inconsistency may happen when the design consideration evolves, and in that case, the initial cDSP template has to be modified. For example, goals, constraints, or parameters are added or removed. Therefore, detecting the inconsistency and

Table 3.9 cDSP template ontology classes [14]

Class	Definition
Problem	Design problem general information, e.g., the product under design, functional requirements to be satisfied, etc
cDSPTemplate	Integrating all template attributes and the related information. Problem formulation. A problem can be formulated as several cDSP templates
Behavior	The template expresses the template's decision model behavioral information, i.e., sensitivity to parameter variation, converging tendency, etc
History	Template history evolution (1) the template from which this template originates and (2) the template developed from this one
Function	General attributes of a constraint and goal
Quantity	General attributes of a parameter and variable
Constraint	A constant value function that cannot be violated. Function Subclass
Goal	A function with a target value that can be violated. Function Subclass
Parameter	A fixed valued quantity during the problem-solving procedure
Variable	A variable valued quantity during the problem-solving procedure
Preference	The designers' preferences considering the system goals' satisfaction
Analysis	The information regarding inputs, outputs, and the related analysis codes
Driver	The problem solvers' interface running the analysis codes
Response	The actual response to a given template specification

Table 3.10 cDSP template ontology data slots [14]

Slot name	Definition	Type
name	Instance name	String
description	Instance description	String
quantityType	*Quantity* types, i.e., "DeviationVariable", "SystemParameter", or "SystemVariable"	Symbol
symbol	A *Quantity's* symbol	String
value	A *Quantity's* value	Float
unit	A *Quantity's* unit	String
lowerBound	A *Quantity's* lower bound	Float
upperbound	A *Quantity's* upper bound	Float
expression	A *Function's* expression	String
functionType	*Function* types, i.e., "SystemGoal" or "SystemConstraint"	Symbol
monotony	*Function* monotony, i.e., "Maximize", "Minimize", or "Force"	Symbol
equality	Constraint *Function* equality ("≤", "≥" or "=")	Symbol
linearity	Function linearity, i.e., "nonlinear" or "linear"	Symbol
target	*Function* target	Float
weight	*Preference* weight in Archimedean form	Float
level	*Preference* level in preemptive form	Symbol
numberOfSamples	Samples cardinality for an explicit *Preference* in the design experiments	Float
problemSolver	Problem solver driving the template	Symbol
codeFileLocation	Storage path of the analysis code file	String
result	*Response* output information	String
behavioralInfo	Template behavioral information	String
modification	Modification information of a template from its predecessor template	String

informing designers to solve it is essential. Similarly to sDSP template ontology, we use a rule-based reasoning method to check the consistency in the cDSP template ontology. The rules to preserve the consistency in the cDSP template ontology are presented in Table 3.12. Considering Rule 6 as an example, the rules are defined as

(defrule MAIN::rule_6 (object (is-a cDSPTemplate) (OBJECT ?y)) => (foreach ?x (slot-get ?y hasParameter) (if (neq (slot-get ?x lowerBound) (slot-get ?x upperBound)) then (printout t WARNING_6 crlf))))

This example means that if any instance in the slot "hasParameter" has unequal values related to the upper and lower bound, the reasoner shall send a message informing the designer working on that cDSP template about the inconsistency.

Table 3.11 cDSP template ontology object slots [14]

Slot name	Definition	Type
elementOf	Relate a *Quantity* to a *Functions* group	Instance
functionOf	Relate *Function* to a *Quantities* group	Instance
associatedGoal	Relate a *Preference* to a *Goal*	Instance
input	Relate as input an *Analysis* to a *Quantities* group	Instance
output	Relate as output an *Analysis* to a *Quantities* group	Instance
hasVariable	Relate a *cDSPTemplate* to a group of *Variables*	Instance
hasParameter	Relate a *cDSPTemplate* to a group of *Parameters*	Instance
hasConstraint	Relate a *cDSPTemplate* to a group of *Constraints*	Instance
hasGoal	Relate a *cDSPTemplate* to a group of *Goals*	Instance
hasPreference	Relate a *Goal* to a *Preference*	Instance
hasDriver	Relate a *cDSPTemplate* to a *Driver*	Instance
hasAnalysis	Relate a *cDSPTemplate* to an *Analysis*	Instance
hasResponse	Relate a *cDSPTemplate* to a *Response*	Instance
hasTemplate	Relate a *Problem* to a *cDSPTemplate* group	Instance
applyTo	Relate a *cDSPTemplate* to a *Problem*	Instance
hasHistory	Relate a *cDSPTemplate* to a *History*	Instance
hasBehavior	Relate a *cDSPTemplate* to a *Behaviors* group	Instance
derivedFrom	Relate a *cDSPTemplate* to another *cDSPTemplate*	Instance
evolveTo	Relate a *cDSPTemplate* to another *cDSPTemplate*	Instance

Table 3.12 cDSP template ontology consistency rules [14]

Rule 1	All instances in slot "hasVariable" must be of the "SystemVariable" type
Rule 2	All instances in slot "hasParameter" must be of the "SystemParameter" type
Rule 3	All instances in slot "hasConstraint" must be of the "SystemConstraint" type
Rule 4	All instances in slot "hasGoal" must be of the "SystemGoal" type
Rule 5	All instances in slot "hasVariable" should conform to lowerBound $\leq$ value $\leq$ upperBound
Rule 6	All instances in slot "hasParameter" should conform to: lowerBound = value = upperBound
Rule 7	All instances in slot "hasPreference" should conform to $0 \leq$ weight ≤ 1
Rule 8	All Variable instances of the "DeviationVariable" type should conform to value ≥ 0
Rule 9	All instances in slot "hasPreference" should conform to: $\sum_0^k w_i = 1$
Rule 10	All Variable instances of the "DeviationVariable" type should conform to: $d_i^- \cdot d_i^+ = 0$
Rule 11	All Variable instances of the "DeviationVariable" type should be included in the "output" slot
Rule 12	All Variable instances of the "SystemVariable" type should be included in the "input" slot

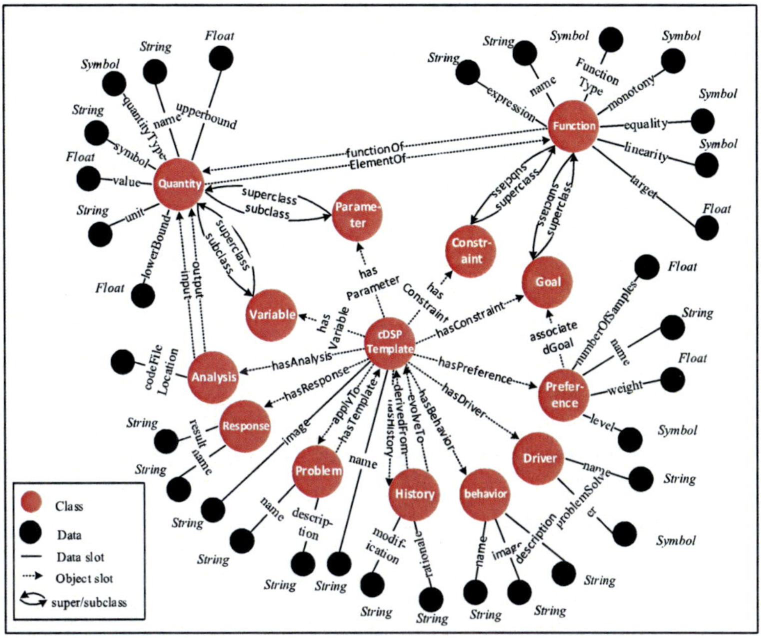

Fig. 3.14 cDSP template ontology architecture [14]

(4) **The Ontology Structure**

The cDSP template ontology structure, illustrated in Fig. 3.14, is a network structure where classes are represented as red nodes and data as black nodes. Object slots, connecting the different classes, are shown with dashed-line arrows, and classes are related to actual data by data slots (solid-line arrows). Hierarchical relations, e.g., the link between Classes *Constraint* and *Function,* are illustrated utilizing arc arrows. The ontology modeling procedure is facilitated using the Protégé tool and a cDSP template instance example is shown in Fig. 3.14. Protégé also makes it possible to include plugins that extend ontology functions. To execute the populated cDSP template instances, we develop a plugin that links DSIDES to Protégé using a Java function call to generate the design solutions.

3.3.4 Test Example—Designing a Pressure Vessel

In this section, we use a thin-walled pressure vessel design example to demonstrate the cDSP template ontology exploitation in terms of facilitating designers making decisions by reusing previous design information. The problem examined here extends the problems used by Lewis and Mistree [18], who illustrated collaboration in decision-based designs. In our example, first, an instance is established employing the initial design decision information. Then, three redesign scenarios are presented for demonstrating the exploitation of the ontological method, and finally, we provide a summary and discussion of the findings.

(1) **Populating an Original Design Instance**

Radius R, length L, and thickness T, are the pressure vessel design variables shown in Fig. 3.15. This vessel is used to withstand a specified internal pressure P utilizing a specific material. The designer has two objectives: minimizing weight and maximizing the cylinder's volume, with both having geometry and stress constraints. For this example, the nomenclature is presented in Table 3.13. The cDSP formulation is given as follows:

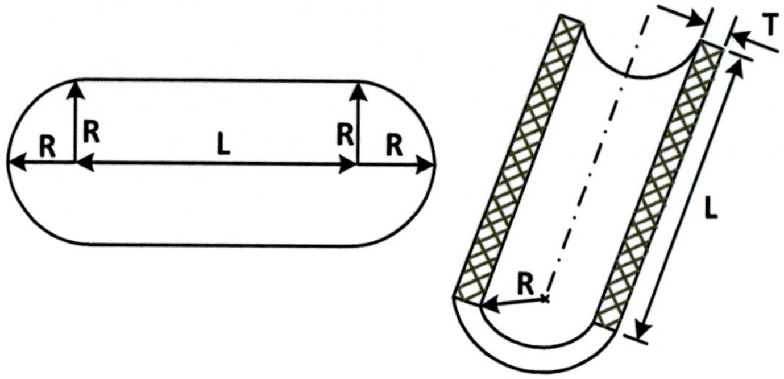

Fig. 3.15 Thin-walled pressure vessel with hemispherical ends [14]

Table 3.13 Pressure vessel example nomenclature [14]

w	Pressure vessel Weight, lbs.
V	Volume, in.^{3}
R	Radius, in.
T	Thickness, in.
L	Length, in.
P	Pressure inside the cylinder, Klb.
S_t	Allowable tensile strength of the cylinder material, Klb.
ρ	Cylinder material density, lbs./in.^{3}
σ_{circ}	Circumferential stress lbs./in.^{2}
TV	Target value for a goal
W_{VOL}	Weight-related with the weight goal
W_{WGT}	Weight-related with the volume goal

Given

- Weight: $W(R,T,L) = \rho[\frac{4}{3}\pi(R+T)^3 + \pi(R+T)^2L - (\frac{4}{3}\pi R^3 + \pi R^2 L)]$
- Volume: $V(R,L) = \frac{4}{3}\pi R^3 + \pi R^2 L$

Find

- System Variables: Radius R, Thickness, T and Length, L
- Overachievement Deviation Variable associated with weight goal, ${d_w}^+$
- Underachievement Deviation Variable associated with volume goal, ${d_v}^-$

Satisfy

- Stress constraint: $c_1 : \sigma_{circ} = \frac{PR}{T} \leq S_t$
- Geometric constraints: $c_2 : 5T - R \leq 0$

$c_3 : R + T - 40 \leq 0$

$c_4 : L + 2R + 2T - 150 \leq 0$

- Bounds: $T_l \leq T \leq T_u$

$R_l \leq R \leq R_u$

$L_l \leq L \leq L_u$

- Weight goal: $W - {d_w}^+ = W_{TV}$
- Volume goal: $V + {d_v}^- = V_{TV}$

Minimize

$W_{VOL}{d_v}^- + W_{WGT}{d_w}^+$

In Table 3.14, we present the input data of this problem. Given all the prepared information, we initiate in Protégé (see Fig. 3.16) a cDSP template instance for the design of the pressure vessel. In this figure, the left-hand side panel refers to the class browser listing all classes; the middle panel represents the instance browser that lists all instances linked with a specific class, while the right-hand side panel refers to the instance editor to create and edit a specific instance. For this scenario, all object and data slots, e.g., "hasVariable", "hasParameter", and "name" of the pressure vessel instance, are created based on the structure defined in Sect. 3.3.3 and the problem's data. For a correctly created instance, the result is calculated using JAVA function calls that communicate with the problem solver DSIDES. The result is *(R, T, L) = (36, 4, 70)*, *(Weight, Volume) = (39,457 lbs., 480,385 in.*^{3}*)* (see front window in Fig. 3.16).

Table 3.14 Pressure vessel input [14]

P	3.89 klb.
S_t	35.0 klb.
ρ	0.283 lbs./in.^{3}
L_l	0.1 in.
L_u	140.0 in.
R_l	0.1 in.
R_u	36.0 in.
T_l	0.5 in.
T_u	6 in.
W_{TV}	0.1 lbs.
V_{TV}	775,000 in.^{3}
W_{VOL}	0.5
W_{WGT}	0.5

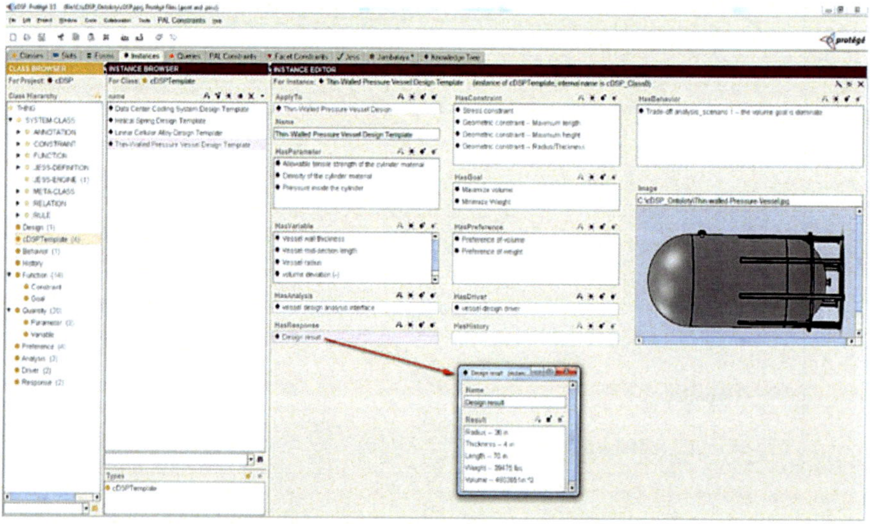

Fig. 3.16 Screenshot of the pressure vessel design cDSP template instance in Protégé ontology editor [14]

(2) **Facilitating Decision-Making in Redesign Scenarios**

Consistency Checking. In this scenario, we assume the original design variable R is fixed to 20 inches due to manufacturing limitations, and therefore, the existing design should be modified to meet the new condition. As shown in Fig. 3.17, this problem relies on a suitable choice of L and T and transforming it from a three-dimensional problem $(L - R - T)$ to a two-dimensional $(L - T)$. In Fig. 3.17, the previously achievable set is a solid vertical box, with one of its ends being the polygon AJIH in the $L = 0$ plane, while the other end is formed by the inclined plane

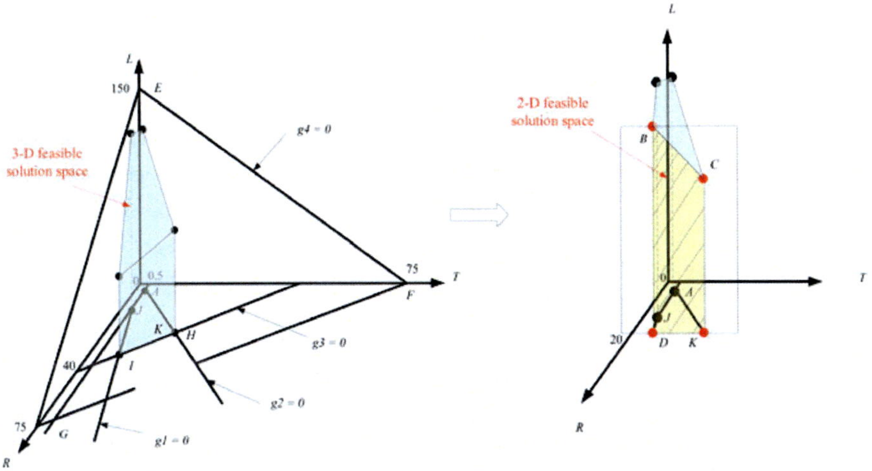

Fig. 3.17 Design space dimension change [14]

passing through the points marked EFG. Introducing the new constraints requires the previously achievable solution space to be sliced by a plane vertical to the R-axis at $R = 20$, and the cross-section marked DBCK forms the new feasible solution space. To solve this problem, designers must adequately adjust the original decision model and form a new decision. The required adjustments are related to **Type I** described in Sect. 3.3.2. In the ontology context, this modification comprises a consistency mechanism shown in Fig. 3.18. The specific processing steps are:

(1) Convert a variable to a parameter—move the previous variable "vessel radius" from slot "hasVariable" to "hasParameter".
(2) Check consistency—check the consistency of the modified cDSP template instance using the JESS rule engine. A message shows, "Rule 6 is violated!—Parameter 'vessel radius' has unequal upper and lower bounds!"
(3) Reconfigure the Parameter—reconfigure the parameter instance "vessel radius" as: *lower bound = value = upper bound = 20 in.*
(4) Obtain the result—problem solver DSIDES calculates and returns the results based on the newly specified cDSP template setup. The result is *(T, L) = (0.5, 53.9)*, *(Weight, Volume) = (1698.7 lbs., 101,191.7 in.*^{3}*)*, which is a new instance.

Trade-off Analysis. In this scenario, we consider setting a new goal aiming to minimize cost in the pressure vessel's design. The cost equation is $C(R, L, T) = 0.6224RTL + 1.7881R^2T + 3.1611T^2L + 19.8621RT^2$, with the target cost being $ 0.1 [19]. The new problem requires adjusting the initial decision model concerning **adding** a new goal and **modifying** the goals' weight value, which is a mixture of modification **Types II** and **III**. We assume that this adjustment is adequately facilitated by the consistency checking mechanism presented in the previous section, and thus here we focus on how trade-off analysis is facilitated in the ontological context.

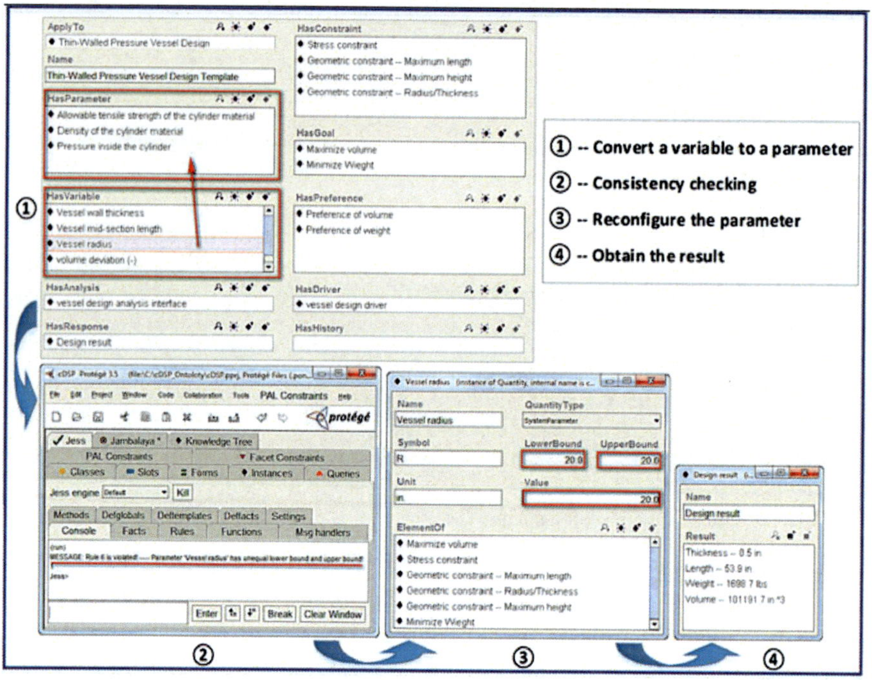

Fig. 3.18 Decision model adjustment facilitated by consistency checking [14]

Trade-off analysis involves analyzing how a different goal priority distribution affects the actual performance of goals, helping designers understand the rationality of their decisions.

Similarly to [20, 21], we employ a ternary plot as a trade-off analysis tool for the three-goal cDSPs. The main idea is to generate a sample of the weighted value sets for the goals and then utilize the latter to calculate each goal's performance (represented as a normalized deviation). Finally, a triangle with a color bar illustrates the relationship between values and performance (Fig. 3.19). Considering the triangle, the edges' scales indicate the goals' weights, the points inside the triangle denote different weight sets, e.g., point *S* refers to set (0.2, 0.4, 0.4), the color refers to the deviation of the associated points, and the color bar gives the corresponding values. For a trade-off analysis, designers must prepare large sample sets and run each sample's computing code in DSIDES to generate the result. Then, all sample sets and the corresponding results are input to Matlab to generate a ternary plot. Typically, this is a time-consuming process that requires automation or platform-related support to improve efficiency.

In the ontology context, the trade-off analysis involves three processes supported by Java function calls and afforded by Protégé, sample generation, communicating to DSDIES for computing, and generating a ternary plot. Sample generation has a critical role in generating quality ternary plots. In the ontology, the data slot "numberOfSamples" (see Table 3.10) controls the sampling procedure by setting the number of

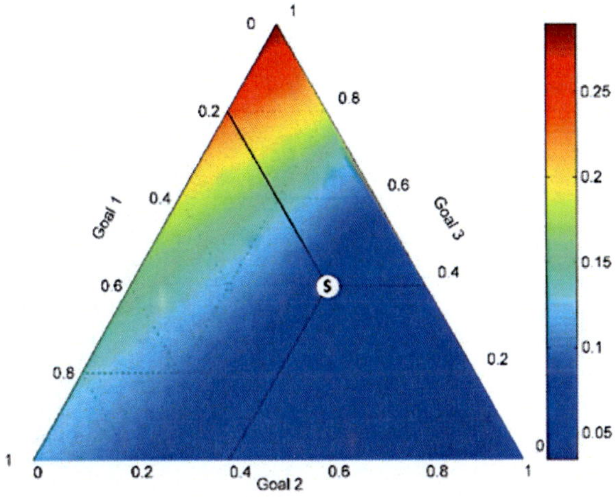

Fig. 3.19 Ternary plot for trade-off analysis

scales for the weight of each goal. This process is executed utilizing a Java function at the computational level, which combines the scales to weight the sets, provided that the scale's sum in each set is one (Fig. 3.20a). In the figure, columns present the goals, the numbers attached to the points in each column indicate the weights' scales, and the colored lines that link the points are the weight sets. In this scenario, we set the "numberOfSamples" slot to three and generate six sample sets ("numberOfSamples" can be increased to generate more sets). Each sample set in Fig. 3.20 is automatically sent to DSIDES to calculate a result. Then, the ternary plot of the new "cost" goal is automatically produced utilizing the sample sets and the results and is captured by the "hasBehavior" object slot (Fig. 3.20b). The blue area of the Ternary plot refers to the preferred weight sets to minimize the goal "cost". The "volume" and the "weight" goal are also plotted for designers' reference.

Design Space Visualization. To visualize the design space, in this scenario, we adopt the assumptions of the previous scenario, i.e., the ternary plot offers designers a trade-off analysis tool concerning the different goals by presenting the relationships between the deviation of the goals and the weight sets. Despite this trade-off analysis assisting designers in making rational decisions, resulting designs may not

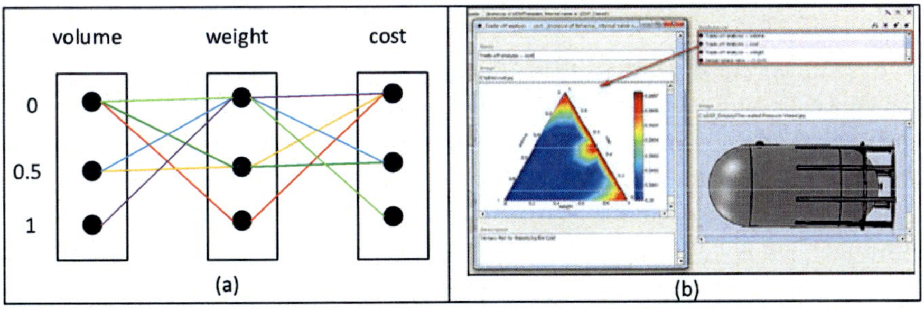

Fig. 3.20 Trade-off analysis in the ontological

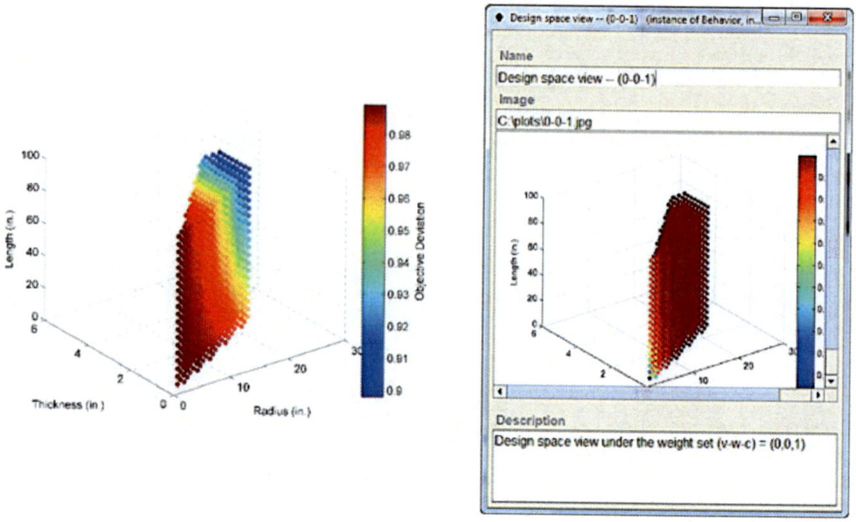

(a) Scatter plot

(b) Scatter plot in protégé

Fig. 3.21 Design space visualization in the ontology

be acceptable. For instance, we set a point ($W_{VOL} = 0$, $W_{WGT} = 0$, $W_{COS} = 1$) that sets a preference for the "cost" goal only within the blue area in the ternary plot. Although designers may choose this point as the priority set to make their decision, the corresponding design ($R = 2.5$, $T = 0.5$, $L = 0.16$) could be unacceptable due to its small size and volume. Therefore, notifying designers about related designs under specific priority sets is crucial to facilitate decision-making, which we realize through utilizing scatter plots[1] (see Fig. 3.21a). In this figure, the axes correspond to the three problem variables, the points correspond to the sample design specifications that are within the achievable design space, and the point's color refers to the objective function deviation of the associated design. Blue points indicate lower deviations and are related to superior designs that the designers are looking for. Hence, in this plot, the blue point locations create an outline of the superior design solutions. For instance, in Fig. 3.21a, for the set point ($W_{VOL} = 0.33$, $W_{WGT} = 0.33$, $W_{COS} = 0.33$), which are acceptable in the ternary plot, the superior designs are located where the associated size is extensive. By visualizing the design space, designers are more confident in their decisions.

In terms of ontology, visualizing the design space is achieved by utilizing Java function calls within Protégé. Such a function gathers information in the "hasParameter", "hasVariable", "hasGoals", "hasConstraints", and "hasPreference" slots and produces a scatter plot showing samples in the feasible design space. Finally, the plot is captured as behavior in the "hasBehavior" slot of a cDSP template instance. The corresponding plot for a priority set (0, 0, 1) is illustrated in Fig. 3.21b, while other

[1] WIKIPEDIA. Link https://en.wikipedia.org/wiki/Scatter_plot refers to Scatter Plot.

priority sets such as (1, 0, 0), (0, 1, 0), (0.33, 0.33, 0.33), etc., are also captured by the "hasBehavior" slot.

(3) **Summary and Discussion**

In this section, based on the thin-walled pressure vessel example, we (i) present a cDSP template instance that exploits the original design decision information, (ii) modify the design variables cardinality to make new decisions that rely on existing template instances, (iii) modify the goals cardinality and the value assigned to weights and goals to make new decisions, (iv) perform a trade-off analysis facilitating decision- making, (V) visualize the design space to enhance decision-making. We verify the ontology-based cDSP template reusability by reusing and modifying previously created instances and prove its executability through consistency checking, automatic results generation, performing trade-off analysis, and visualizing the design space. The results highlight that the designers' decision-making process is facilitated concerning:

- Decision model editing (cDSP template instance) ensuring consistency.
- Fast decision-making assisted by automatic execution schemes.
- Rational decision-making is assisted by insightful information display (ternary and scatter plots).

3.4 Ontology-Based Representation of Coupled Hierarchical Decisions

In this section, we present an ontology for representing coupled hierarchical decisions. First, we show the mathematical model of the coupled hierarchical decisions, and second, we identify the requirements for modeling the knowledge related to coupled hierarchical decisions. Third, we develop an ontology using Protégé for addressing the requirements, and finally, we use a portal frame design example to test the ontology's efficiency.

3.4.1 Mathematical Model for Coupled Hierarchical Decisions

A coupled hierarchical decision involves the solution of any combination of selection and/or compromise DSPs simultaneously by reformulating all the DSPs into a single cDSP. Publications on coupled DSPs include coupled compromise-compromise DSPs [22, 23], coupled selection-compromise DSPs [24, 25] and coupled selection-selection DSPs [26]. The mathematical formulation of the coupled selection-compromise DSP is shown in Fig. 3.22, while the other two coupling types can be formulated similarly. In Fig. 3.22 X and Y denote the variables of two decisions.

Given

Selection

M	Alternatives
N	Attributes
R_{ij}	Normalized rating of alternative i concerning attribute j
I_j	Relative importance of attribute j
MF_i	Merit function for the ith alternative where $MF_i = \sum_{j=1}^{N} I_j R_{ij}$ $(i = 1, \dots, M)$

Compromise

n	Number of system variables
p	Number of equality system constraints
q	Number of inequality system constraints
$p+q$	Total number of system constraints
m	Number of system goals (selection plus compromise)
m_1	Number of coupled compromise system goals
$g_i(X,Y)$	Compromise constraint function
$A_i(X,Y)$	Compromise achievement function
G_i	Target of compromise goal

Find

Selection Variables

$X_i,\ e_i^-,\ e_i^+$ $(i = 1, \dots, M)$

Compromise Variables

Y_i $(i = 1, \dots, n)$

$d_i^-,\ d_i^+$ $(i = 1, \dots, m)$

Satisfy

Selection System Constraint:

$\sum_{i=1}^{N} X_i = 1$

Selection System Goals

$MF_i \cdot X_i + e_i^- - e_i^+ = 1$ $(i = 1, \dots, M)$

Compromise System Constraints

$g_i(X,Y) = 0$ $(i = 1, \dots, p)$

$g_i(X,Y) \geq 0$ $(i = p+1, \dots, p+q)$

Compromise System Goals

Coupled goals

$A_i(X,Y) + d_i^- - d_i^+ = G_i$ $(i = 1, \dots, m_1)$

Independent goals

$A_i(Y) + d_i^- - d_i^+ = G_i$ $(i = m_1 + 1, \dots, m)$

Bounds

$0 \leq X_i \leq 1$ $(i = 1, \dots, M)$

$Y_j^{min} \leq Y_j \leq Y_j^{max}$ $(j = 1, \dots, n)$

$e_i^-,\ e_i^+ \geq 0 \text{ and } e_i^- \cdot e_i^+ = 0$ $(i = 1, \dots, M)$

$d_i^-,\ d_i^+ \geq 0 \text{ and } d_i^- \cdot d_i^+ = 0$ $(i = 1, \dots, m)$

Minimize

The Deviation Function (Preemptive form):

$Z = [f_q(d_i^-, d_i^+), \dots, f_1(e_i^-, e_i^+)] \text{or} [f_1(e_i^-, e_i^+), \dots, f_q(d_i^-, d_i^+)]$

Fig. 3.22 Mathematical formulation of the coupled selection-compromise DSP

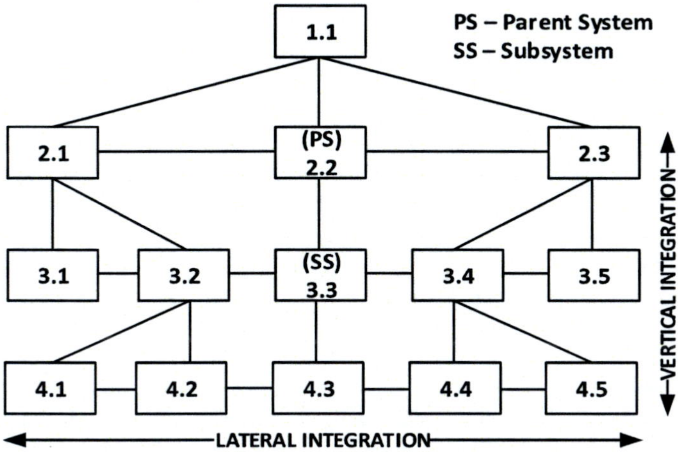

Fig. 3.23 A general four-level hierarchical system [29]

System constraints and goals (e.g., *MF(Y)X, g(X, Y)* and *A(X, Y)*) that involve both *X* and *Y* represent the lateral interactions among member decisions in a hierarchy. Vertical interactions (not shown in the figure) can also be modeled with system goals as $X = X(Y)$, where *X* represents the parent decision variables, and *Y* the sub-decision variables. For details; see [23, 27, 28]. Coupled DSPs are generally multi-objective, nonlinear, mixed discrete–continuous problems. Such problems can be solved using DSIDES combined with the ALP algorithm [17].

A general representation (four-level) for a hierarchical system is proposed by Sobieszczanski-Sobieski [29], as shown in Fig. 3.23. The nature of the hierarchy is a two-dimensional (horizontal and vertical) network structure. The nodes represent the integral parts of the system that can be parent systems or subsystems depending on their relative positions in the hierarchy. The links represent vertical and lateral dependencies between two nodes, depending on whether they cross their levels. In the vertical direction, the root node, e.g., Node "1.1" in Fig. 3.23, is progressively decomposed into multiple levels, and each level may have multiple nodes. Every two levels in the neighborhood are interconnected by a vertical dependency between the nodes of the two levels, which must be subject to the relationship of "parent–child". Thus, the network is vertically integrated. In the horizontal direction, nodes of the same (e.g., Nodes "3.1" and "3.2") or different (e.g., Nodes "3.2" and "3.3") parent(s) at the same level may be connected by lateral dependencies. Thus, the network is laterally integrated.

3.4.2 *Requirements for Knowledge Modeling to Support Hierarchical Decisions*

From the decision-based design perspective, the design of such hierarchical systems is supported by the DSP networks in which each node embodies a DSP that supports the design decision about the related system element, and each link embodies the interaction between two nodes. To facilitate design engineers in formulating a decision hierarchy and solving the problem in keeping with the procedures of hierarchical system designs, we identify the following requirements for the computational DSP hierarchy model:

- **Decomposability**—The multilevel or multiscale characteristic of hierarchical systems requires the model to be decomposable to capture the complexity of the physical system and be solvable using the available computational power. Furthermore, the model should support dynamic decomposition as the design evolves.
- **Flexibility**—The evolution of knowledge about a system being designed at different design stages requires the model to be flexible enough to support reconfiguration. For example, in the early stages, the system is not decomposed into very detailed levels because of the lack of knowledge, and much of the dependency is ignored. As time moves forward, knowledge increases and the original model must be reconfigured to incorporate this knowledge.
- **Visualization—**A comprehensive understanding of the design problem requires the model to be visualized. A hierarchical system is a system of parent systems and interacting subsystems. The computational model should support visualization of the hierarchical structure to facilitate designs' intuitive understanding of the problem they are dealing with and edit the model effectively.
- **Reusability**—Same as the definition of reusability for a cDSP (see Sect. 3.3.1).
- **Executability**—Same as the definition of executability for a cDSP (see Sect. 3.3.1).
- **Consistency—**Same as the definition of consistency for a cDSP (see Sect. 3.3.1).

3.4.3 *Ontology Development for Decision Hierarchies*

In this section, we present an ontology for the computational model (template) of a DSP hierarchy considering the requirements identified in Sect. 3.4.2. The utility of an ontology in facilitating the reuse, execution, and consistency maintenance of the DSP templates has been illustrated in Sects. 3.2 and 3.3. In particular, the reuse of the DSP templates is facilitated with a common vocabulary embodied in the ontology for generating different specific instances that can easily be adapted to new scenarios by modifying the associated Slots. The execution of the DSP templates is facilitated with OWL—a standard and computer-interpretable modeling language underlying the ontology, which supports the parsing by other applications through, for example, Java function calls. The consistency of the DSP templates is maintained by incorporating reasoning mechanisms for consistency checking within the ontology. In this section, the ontology is extended to model a hierarchical DSP network, emphasizing to meet the decomposability, flexibility, and visualization requirements listed in Sect. 3.4.2.

In a computational environment, the decomposability of a model means that it can be further divided or broken down into smaller components. In a DSP hierarchy, this involves vertical integration of dependent sub-DSPs. This calls for the identification of a concept from a higher level than the DSPs alone. This is a general representation of a hierarchy that incorporates the DSPs and captures the associated links so that newly derived sub-DSPs can be linked to an existing hierarchy when decomposition or reassembly is needed. In this paper, the concept is named ***Process***, which is formally defined as an ontology Class. The flexibility of the DSP hierarchy is embodied in reconfiguration, which in principle includes (I) dynamic decomposition of decisions and (II) incorporation of evolving dependencies (or links) among decisions. This requires the links in the hierarchy to be separately modeled so that they can be dynamically added, edited, and removed when reconfiguration is needed. The links are modeled as ***Interfaces***, which are discussed later. Visualization is implemented with a graph-based editing tool, which is presented later.

(1) Definition of Class *Process*

The Class ***Process*** is a general representation of the building blocks in a hierarchy, which incorporates one DSP and its associated dependency. For example, in Fig. 3.23, the collection of Node 2.2 and its associated links at the top, bottom, left-hand side, and right-hand side can be called an Instance of a ***Process***. It is called ***Process*** because it represents an information processing unit based on a decision-making mechanism (sDSP or cDSP) and the associated interaction (information flows) with other units. The concept of the Class ***Process*** is shown in Fig. 3.24. It represents a standard, scalable hierarchy building block where the solid box (or shell) refers to the information processing unit with a DSP (the dashed box) plugged in, and lines stand for the associated vertical and lateral dependencies. A hierarchy is built by assembling a series of different ***Processes***. In the ontology context, the Slots of the Class ***Process*** are defined in Table 3.15.

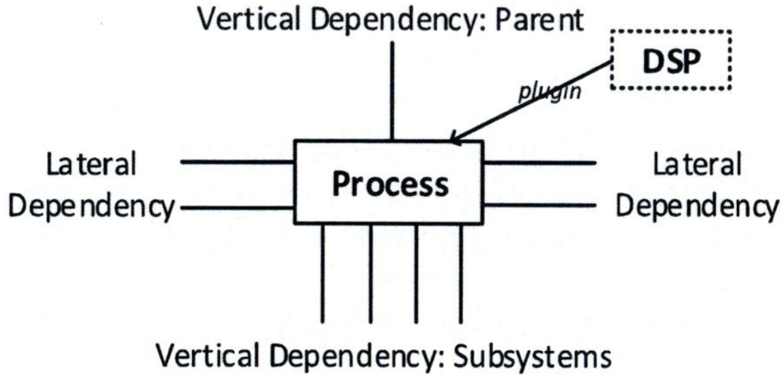

Fig. 3.24 Concept of class process

Table 3.15 Slots of class process

Slot name	Definition	Type
name	***Process*** name	String
description	***Process*** description	String
IsRoot	Indicator of whether the ***Process*** is the root (i.e., no parent) of the hierarchy. Two allowable values—"Yes (1)" or "No (0)"	Boolean
IsLeaf	Indicator of whether the ***Process*** is the leaf (i.e., no subsystems) of the hierarchy. Two allowable values—"Yes (1)" or "No (0)"	Boolean
Decision	Decision corresponding to the ***Process***. Two allowable Classes, namely, cDSP Template (see [30]) and sDSP Template (see [31]). One ***Processes*** Instance can have only one corresponding decision, selection, or compromise	Instance
DecisionType	Indicator of the type of the corresponding decision. Two allowable values— "Compromise" or "Selection"	Symbol
LateralDependency	Lateral dependencies of the ***Process***. It is associated with the Class ***Interface*** introduced in the next section. One ***Processes*** Instance can have multiple lateral dependencies	Instance
VerticalDependency_Parent	Parent dependency of the ***Process***. It is associated with the Class ***Interface*** introduced in the next section. One ***Processes*** Instance can have only one parent dependency	Instance
VerticalDependency_Subsystem	Subsystem dependency of the ***Process***. It is associated with the Class ***Interface*** introduced in the next section. One ***Processes*** Instance can have multiple subsystem dependencies	Instance

(2) **Definition of Class *Interface***

The Class ***Interface*** is a representation of the vertical or lateral dependency between two different ***Processes***. For example, in Fig. 3.23, the link between Nodes 2.2 and 3.3 can be called an instance of an ***Interface***. It is called an Interface because it captures the communication between the two processes. ***Interfaces*** are critical in building a DSP hierarchy because they constitute the medium connecting the individual building blocks, namely, ***Processes***, to an integrated whole. The concept of ***Interface*** is shown in Fig. 3.25, which highlights that an ***Interface*** consists of two elements: the references (or indices) of two linked ***Processes*** represented by dashed boxes, and the information flow, which is represented by the solid line between the two ***Processes***. Based on the strength, the information flow is categorized into two types: weak and strong flow. The weak flow is a one-way flow ("1 to 2" or "2 to 1"), which means that one ***Process*** has parameters that need to be input from the counterpart.

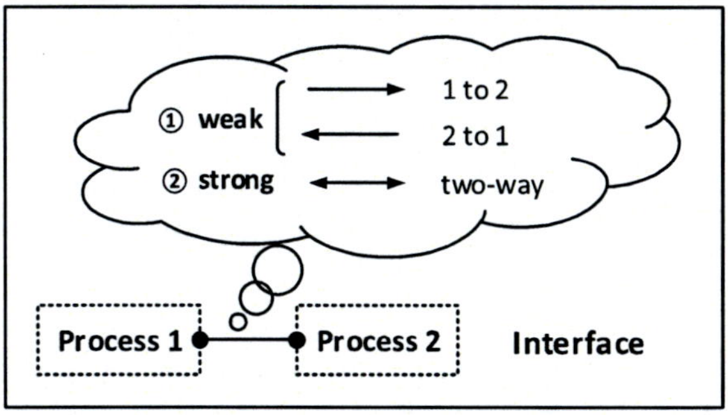

Fig. 3.25 Concept of class interface

A strong flow is a two-way flow, meaning both ***processes*** have parameters that need to be input from the counterpart.

Based on the direction, information is divided into two types, i.e., lateral and vertical flows. Lateral flow is a flow that links the ***Processes*** at the same level in a hierarchy. Vertical flow is a flow that links Processes at different (neighboring) levels in a hierarchy. Since the ***Processes*** are embedded with DSPs, the information flow must be modeled to be consistent with the coupled DSP construct introduced in Sect. 3.4.1. Using the interface shown in Fig. 3.25 as an example, the information flow characteristics in the context of coupled DSP construct are modeled as follows. Here, we assume X_1 represents the variable vector of the DSP embedded in ***Process*** 1 and X_2 the DSP embedded in ***Process*** 2.

- **Vertical—**The vertical information flow is usually two-way and occurs between cDSPs in a hierarchy. Assuming ***Process*** 1 is the parent system and ***Process*** 2 the subsystem, the coupling is modeled as a system goal of a coupled cDSP, formulated as $A(X_1, X_2) = \frac{X_1}{X_1(X_2)} - 1$, where $X_1(X_2)$ is a function that maps X_2 to X_1.
- **Lateral—**In this type, the coupling is divided into the following six sub-types:
  - **sDSP↔sDSP**—Two-way flow between two sDSPs. This can be embodied as: dependent attributes, dependent alternatives, dependent alternatives, and attributes. For further details, the reader is referred to [32]. Mathematically, the interdependency is modeled as a system goal of a coupled cDSP.
  - **sDSP→cDSP—**One-way flow from sDSP to cDSP. Assuming that ***Process*** 1 is embedded with an sDSP and ***Process*** 2 with a cDSP, the information flow is embodied by the Boolean-type variable vector X_1 of the sDSP that constitutes the parameter of the cDSP's system constraints represented as $g(X_1, X_2)$, and (or) the system goals as $A(X_1, X_2)$.
  - **cDSP→sDSP—**One-way flow from cDSP to sDSP. Assuming that ***Process*** 1 is embedded with a cDSP and ***Process*** 2 with an sDSP, then the information flow is embodied by the variable vector X_1 of the cDSP that constitutes the parameter of the sDSP's merit function represented as $MF(X_1) \cdot X_2$. The latter is a system goal in a coupled cDSP. Here, X_2 is Boolean.

- **sDSP↔cDSP**—Two-way flow between an sDSP and a cDSP. This is embodied by the combination of the preceding two types.
- **cDSP→cDSP**—One-way flow between two cDSPs. This is embodied by the variable vector of the antecedent cDSP that constitutes the system constraints and goals of the subsequent cDSP.
- **cDSP↔cDSP**—Two-way flow between two cDSPs. This is embodied by the variable vectors of the two cDSPs that constitute the counterpart's system constraints and goals.

The resolution of a coupled DSP involves reformulating all DSPs into one single cDSP. In this paper, since the ***Interface*** constitutes the dependent part (coupled system constraints and goals) of the DSPs, it should be integrated with the independent parts (embedded in the ***Processes***) to compose a single cDSP. To represent these dependencies in a hierarchy, in the ontology, we identify and define the Slots of Class ***Interface*** in Table 3.16.

(3) **Building DSP Hierarchies Using *Processes* and *Interfaces***

The definitions of Class ***Process***, Class ***Interface,*** and the associated Slots are facilitated by using the Protégé 3.5 tool, as shown in Fig. 3.26. The panel marked with "①" is the class browser where ontology Classes list Classes of the cDSP ontology (termed "CO" [30]), Classes of the sDSP ontology (termed "SO" [31]), and the two Classes (highlighted in the box) identified in this paper for building DSP hierarchies. The window marked with "②" is the Instance editor for the ***Process*** Instances, where the Slots are populated with specific problem information. The window marked with

Table 3.16 Slots of class interface

Slot name	Definition	Type
name	***Interface*** name	String
description	***Interface*** description	String
snterfaceType	Indicator of the type of the ***Interface***. Two allowable values—"Vertical" or "Lateral"	Symbol
strength	Indicator of the strength of the coupling. Two allowable values—"Weak" or "Strong"	Symbol
originalProcess	**The original *Process* is linked by the *Interface*.** It is associated with the Class ***Process***	Instance
counterpartProcess	The counterpart ***Process*** is linked by the ***Interface***. It is associated with the Class ***Process***	Symbol
originalCounterpartFlow	Information flow from the original ***Process*** to its counterpart. It is associated with Class ***Function*** (see [30] for detailed definition). One ***Interface*** Instance can have multiple flows from 1 to 2	Instance
counterpartOriginalFlow	Information flow from the counterpart ***Process*** to the original ***Process***. The definition is similar to Slot "Flow:1_to_2"	Instance

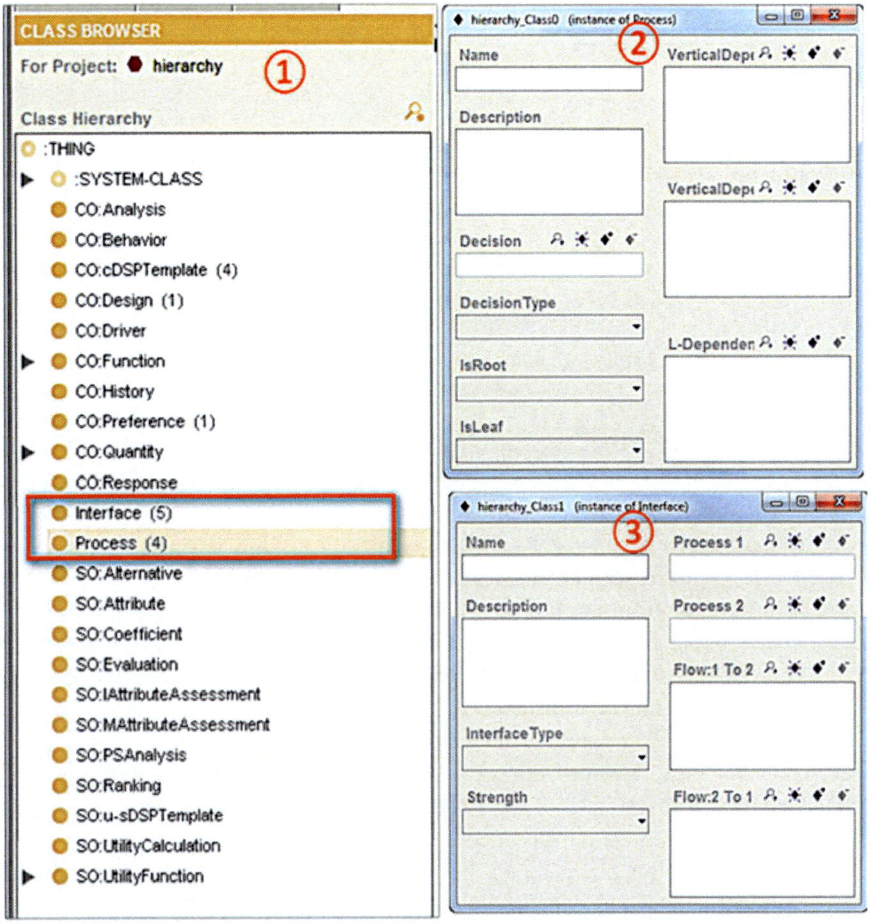

Fig. 3.26 User-interfaces of the DSP hierarchy ontology [14]

"③" is the Instance editor for the ***Interface*** Instances, where the Slots are populated using specific problem information.

The construction of DSP hierarchies is facilitated by the Protégé Graph Widget [33], a graphical tool for visually editing the Instance and the relationships among Instances. The Protégé Graph Widget is especially suitable for building the DSP hierarchy, as the latter is a network of ***Process*** Instances and ***Interface*** Instance. A screenshot of the widget customized for building the hierarchy is shown in Fig. 3.27. The corresponding building procedure is the following:

Step 1. Creating ***Process*** Instances. This is done by first dragging the box marked "Process" from the left panel (which represents Classes) to the right canvas (which represents Instances) and then editing the generated Instance using editor "②" in Fig. 3.26. The number of ***Process*** Instances is determined by the number of decisions made in a practical problem. In Fig. 3.27, "A" and "B" are two ***Process*** Instances.

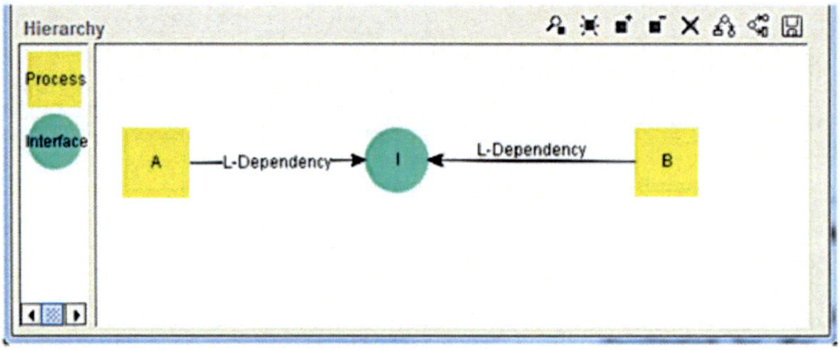

Fig. 3.27 DSP hierarchy in Protégé graph widget [14]

Step 2. Embedding DSPs into each ***Process*** Instance. This is done by first creating the DSP (cDSP and sDSP) Instances using the specific information of the problem in the cDSP and sDSP template editors (see [30, 31]), then embedding the created DSP Instances into each ***Process*** Instance by specifying Slot "Decision".

Step 3. Creating ***Interface*** Instances. This is done first by dragging the circle marked "Interface" from the left panel to the right canvas, then editing the generated Instance using editor "②" in Fig. 3.26. The number of ***Interface*** Instances is determined by the number of dependencies among the decisions in a problem. In Fig. 3.27, "I" is an ***Interface*** Instance.

Step 4. Linking the ***Process*** Instances and ***Interface*** Instances. This is done by specifying the Slots "LateralDependency", "VerticalDependency_Parent", "VerticalDependency_Subsystems" of the ***Process*** Instances, and Slots "Process_1" and "Process_1" of the ***Interface*** Instances. The links are automatically shown in the canvas when these slots are specified. In Fig. 3.27, ***Interface*** Instance "I" represents the lateral interactions between ***Process*** Instances "A" and "B".

Step 5. Run the model and solve the problem in DSIDES. This is done utilizing the Java function calls in Protégé, which integrates all the information specified in the hierarchy as a coupled cDSP and then sends it to DSIDES for computing, solving the problem, and finally obtaining the results from DSIDES.

3.4.4 Test Example—Designing a Portal Frame

In this section, we illustrate the decision hierarchy ontology validity utilizing as an example a portal frame design problem. This example extends the problem of Sobieski [29] to demonstrate the decomposition method and the problem considered by Shupe et al. [23], Allen et al. [27], and Vadde et al. [28] to demonstrate the advantages of the cDSP in solving hierarchical design problems. The correctness of the ontology is tested utilizing a refinement process of the decision model (DSP) corresponding to the portal frame design.

(1) Creation of a Baseline Model with Limited Information

A portal frame represents a simple hierarchical system, as shown in Fig. 3.28. The integrated frame represents the parent system, while the three I-beams are the subsystems. The design objective is to minimize the overall mass of the frame. The frame is subject to two types of constraints: external and internal constraints, with the former including the static loads P and M, while the latter include normal stress, bending stress, shear stress, and buckling in each member. Design variables are categorized into parent system variables and subsystem variables. Parent system design variables A and I stand for each member's cross-sectional area and moment of inertia, respectively. Subsystem variables are the dimensions, namely, b_1, b_2, h, t_1, t_2, t_3 of each subsystem. V_i denotes the vertical interactions between the parent system and each of its subsystems and L_{ij} the lateral interactions between subsystems.

At an early design stage, designers usually face the challenge of limited information for modeling the problem. In the design of the portal frame, we assume that information of the vertical interactions V_i and the lateral interactions L_{ij} are unknown to designers. Designers are required to create a baseline model to design the portal frame with limited information. The baseline model is a single cDSP in which all the

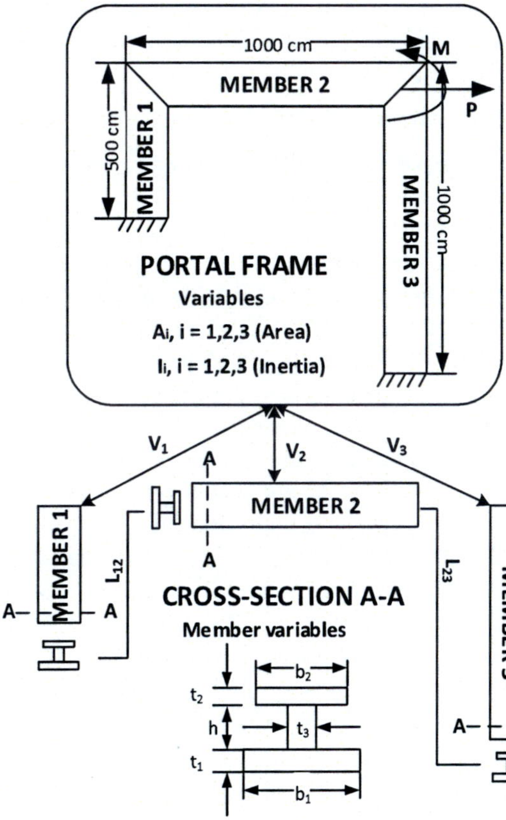

Fig. 3.28 Design of a portal frame [14]

constraints are in terms of the design variables at the subsystem level. In the context of Fig. 3.28, it involves only subsystem variables b, t, and h for each beam and does not exploit the parent variables A and I. The mathematical formulation of this single cDSP is shown in Fig. 3.29. In the ontological context, a ***Process*** Instance marked as "Portal Frame Design" (see Fig. 3.30) is created using the information presented in Fig. 3.29. Since the vertical and lateral interactions are not modeled, there is no ***Interface*** Instance created. Specification of the Process Instance is presented in window "①" of Fig. 3.30, and the information of the cDSP template embedded in the Instance is presented in window "②". Given $L_1 = 500$ cm, $L_2 = 1000$ cm, $L_3 = 1000$ cm, $P = 50,000$ N, $M = 20 \times 10^6$ N cm, the results corresponding to the baseline model are shown in Table 3.17. In Table 3.17, due to not considering the interactions among the three members (subsystems) of the portal frame, there is a noticeable difference between the resulting dimensions of the members. Using the baseline model as the starting point, it is assumed that designers will gradually refine the model by considering lateral and vertical interactions. The performance of the ontology in terms of facilitating the creation of a DSP hierarchy to support this refinement is tested in the remainder of this section. Designers refine the baseline model by decomposing the model into sub-DSPs and then linking the sub-DSPs using lateral interaction information. They further refine the laterally interacted model by incorporating vertical interactions with a parent DSP that supports the decision-making about the parent system. Finally, the model becomes a comprehensive model with both lateral and vertical interactions.

(2) Refinement of the Model with Lateral Interactions

Based on the baseline model, designers are assumed to do the refinement with known lateral interactions between Member 1 and Member 2 and the lateral interactions between Member 2 and Member 3, namely, L_{12}, and L_{23} in Fig. 3.28. This means that the baseline model must be decomposed into three DSPs, and the associated dependency needs to be modeled. The model can be decomposed by separating the cDSP formulation into three independent cDSPs, just by setting index i with a value of 1, 2, and 3, respectively. The (lateral) dependencies necessitate constraints that match the subsystem variables to their counterparts in the other subsystems. This "matching" is modeled mathematically by system goals in a cDSP. For example, in Fig. 3.28, the individual dimensions of the center beam (Member 2) should match those of the beams (Members 1 and 3) on either side. Thus, lateral equality constraints are created to handle this, for example, $b_{1,1} + d^- - d^+ = b_{2,1}$, where $b_{1,1}$ and $b_{2,1}$ are the width of the bottom flanges of members 1 and 2, respectively. The deviation from this equality is measured by the system goal's deviation variables (underachievement d^- and overachievement d^+). Given that the dimensions of the three subsystems should match, lateral interactions are identified and modeled as system goals as follows, where m stands for the lateral interaction index ($m = 1$ is the first lateral interaction, and $m = 2$ the second).

$$b_{1,m} + d_m^- - d_m^+ = b_{1,m+1}, \quad m = 1, 2.$$
$$b_{(2,m)} + d_{(2+m)}^- - d_{(2+m)}^+ = b_{(2,m+1)}, \quad m = 1, 2.$$

Given

b, t, h	dimension variables determined at subsystem level.
E	Young's modulus of the material.
V^E	expected value for mass.
$S(b, t, h)$	combined normal and bending stress for member *i*.
L^b, L^t, L^n	minimum permissible values for *b, t, and h.*
U^b, U^t, U^n	maximum permissible values for *b, t, and h.*
i	1,2,3; member number.
j	1- indicates the left end of member *i*; 2- indicates the right end of member *i*.
σ_a	allowable normal stress on yielding.
σ_{ab}	allowable normal stress on buckling.
τ_a	allowable shear stress on yielding.
τ_{ab}	allowable shear stress on buckling.

Find

b_{1i}, b_{2i}	flange width.
t_{1i}, t_{2i}	flange thickness.
t_{3i}	web thickness
h_i	height of the web.
d^-, d^+	deviation variables.

Satisfy

System constraint

- Combined stress constraints
  $S(b, t, h) \leq S_{max},$
- Combined stress in the top flange
  $\sigma_a/\sigma(b, t, h)_{ij} \geq 1.0,$ $\quad i = 1,2,3; j = 1,2.$
- Combined stress in the bottom flange
  $\sigma_a/\sigma(b, t, h)_{ij} \geq 1.0,$ $\quad i = 1,2,3; j = 1,2.$
- Shear stress constraint
  $\tau_a/\tau(b, t, h)_{ij} \geq 1.0,$ $\quad i = 1,2,3; j = 1,2.$
- Flange buckling constraint
- Normal stress
  $\sigma_{ab}(b_{1i}, t_{1i})_j/|\sigma(b, t, h)_{ij}| \geq 1.0,$ $\quad i = 1,2,3; j = 1,2.$
  $\sigma_{ab}(b_{2i}, t_{2i})_j/|\sigma(b, t, h)_{ij}| \geq 1.0,$ $\quad i = 1,2,3; j = 1,2.$
- Shear buckling stress constraint of the web
  $\tau_{ab}(t_{2i}, h_i)_j/|\tau(b, t, h)_{ij}| \geq 1.0,$ $\quad i = 1,2,3; j = 1,2.$

System goals

- Mass goal: minimize the mass of the system
  $V(b, t, h)/V^E + d^- - d^+ = 1.0,$

Bounds on subsystem variables

$L^b \leq b \leq U^b.$
$L^t \leq t \leq U^t.$
$L^h \leq h \leq U^h.$

Minimize

The Deviation Function (Archimedean form)

$Z = w_1 d^- + w_2 d^+$

Fig. 3.29 A cDSP formulation at subsystem level with no interaction

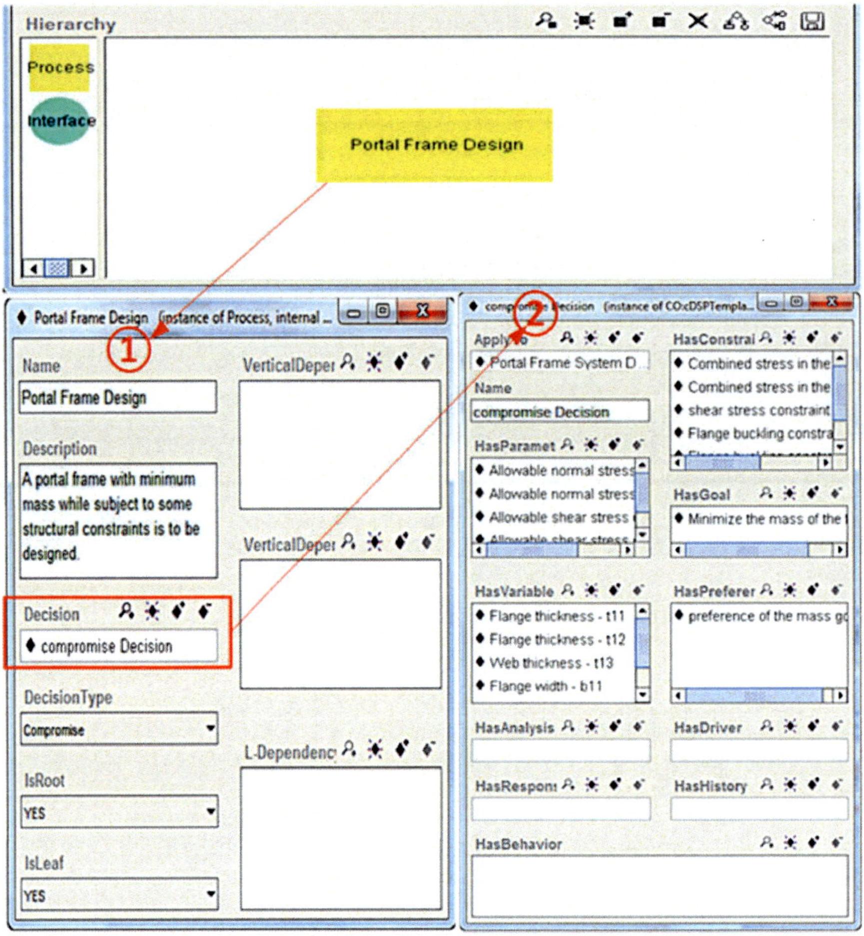

Fig. 3.30 Specification of the model with no interaction included [14]

Table 3.17 Results of the baseline model [14]

Member 1 (cm)		Member 2 (cm)		Member 3 (cm)	
Variable	Value	Variable	Value	Variable	Value
b_1	12.38	b_1	10.97	b_1	11.86
b_2	12.33	b_2	13.11	b_2	11.08
t_1	0.531	t_1	0.492	t_1	0.526
t_2	0.641	t_2	0.579	t_2	0.491
t_3	0.450	t_3	0.313	t_3	0.282
h	62.81	h	59.5	h	52.84
Volume (cm^3): 79,537					

$$t_{(1,m)} + d^-_{(4+m)} - d^+_{(4+m)} = t_{(1,m+1)}, \quad m = 1, 2.$$
$$t_{(2,m)} + d^-_{(6+m)} - d^+_{(6+m)} = t_{(2,m+1)}, \quad m = 1, 2.$$
$$t_{(3,m)} + d^-_{(8+m)} - d^+_{(8+m)} = t_{(3,m+1)}, \quad m = 1, 2.$$
$$h_m + d^-_{(10+m)} - d^+_{(10+m)} = h_{(m+1)}, \quad m = 1, 2.$$

In the ontology, system goals are captured and used to populate the two ***Interface*** Instances, namely, "L12" and "L23", as shown in the canvas of Fig. 3.31. The three separated cDSPs are used to instantiate three ***Process*** Instances, marked "M1", "M2," and "M3", which are linked by "L12" and "L23" that represent strong coupling with two-way information flows. The information flows are embodied in Slots "Flow_1_to_2" and "Flow_2_to_1", as shown in the window under the canvas in Fig. 3.31, in which the specification of "L12" is shown. The deviation variables associated with the interaction system goals and the deviation associated with the

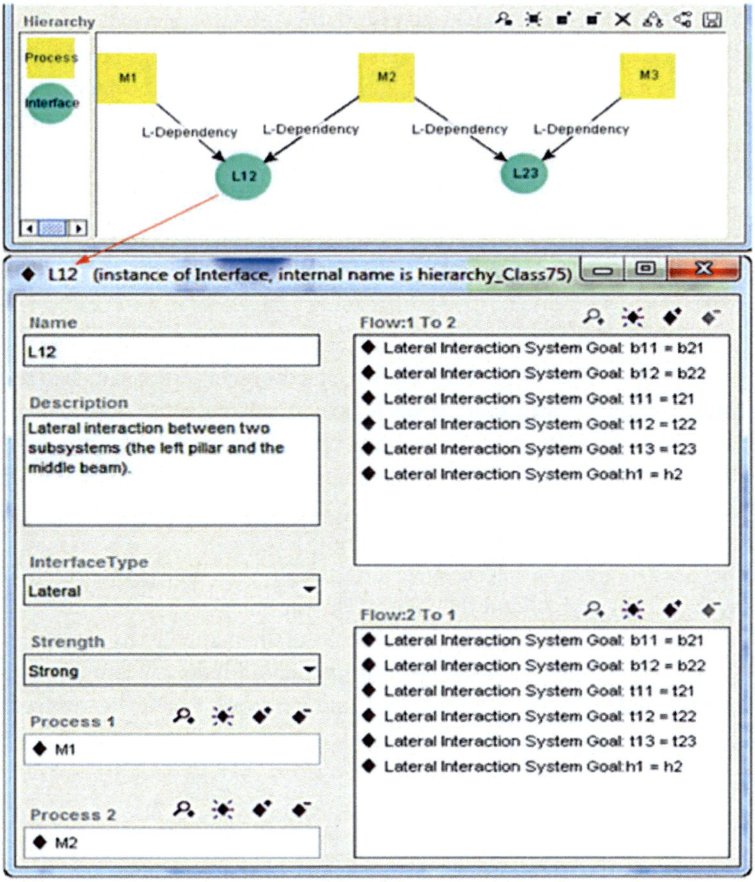

Fig. 3.31 Specification of the model with only lateral interactions included [14]

Table 3.18 Results of the model with only lateral interactions [14]

Member 1 (cm)		Member 2 (cm)		Member 3 (cm)	
Variable	Value	Variable	Value	Variable	Value
b_1	14.43	b_1	14.42	b_1	14.42
b_2	15.72	b_2	15.72	b_2	15.71
t_1	0.725	t_1	0.728	t_1	0.730
t_2	0.735	t_2	0.737	t_2	0.739
t_3	0.409	t_3	0.409	t_3	0.347
h	50.24	h	50.21	h	36.10
Volume (cm^3): 98,605					

mass goal are formulated in a preemptive form in the overall deviation function. The interaction system goals are of higher priority than the mass goal. The overall deviation function is captured in one of the three ***Process*** Instances (in this case is "M1"). All the Process and Interface Instances information are integrated as a coupled cDSP and sent to DSIDES for computation at a computational level. The corresponding results are shown in Table 3.18, indicating that the difference between the subsystem dimensions is reduced to a tolerable extent because of the lateral interactions that match the three subsystems. However, the volume increases from 79,537 to 98,605 cm^3. The difference is due to the low priority of the mass goal in the deviation function.

(3) A Comprehensive Model with Lateral Interactions and Vertical Interactions

In this section, the vertical interactions between the parent system and subsystems, namely, V_1, V_2, and V_3 in Fig. 3.28 are known, and designers are assumed to refine the model further using this information. Since the vertical interactions are known, a new cDSP corresponding to the parent system is created so that the existing subsystem DSP is connected to it through the vertical interactions. The new parent cDSP involves parent system-level design variables A_i and I_i ($i = 1, 2, 3$), constraints and goals, as shown in the mathematical formulation in Fig. 3.32.

The vertical interactions necessitate constraints that match the parent system design variables (A and I) and the subsystem variables (b, t, h). Similarly, to the lateral interactions, this "matching" is modeled mathematically by system goals in a cDSP as follows (i stands for the vertical interaction index, where $i = 1$ denotes the first vertical interaction V_1, while $i = 2$ and $i = 3$ are V_2 and V_3, respectively):

$$A_i + d_i^- - d_i^+ = A(b, t, h)_i, \quad i = 1, 2, 3.$$
$$I_i + d_{i+3}^- - d_{i+3}^+ = I(b, t, h)_i, \quad i = 1, 2, 3.$$

In the ontology, vertical interactions are captured and used to populate ***Interface*** Instances, namely, "V1", "V2", and "V3", as shown in Fig. 3.33. The information of the cDSP corresponding to the parent system is used to instantiate a ***Process*** Instance

Given

E Young's modulus of the material.

V^E expected value for mass.

$S(A_i, I_i)$ combined normal and bending stress for member *i*.

L^A, L^I minimum permissible values for A_i and L_i.

U^A, U^I maximum permissible values for A_i and L_i.

i 1,2,3; member number.

Find

A_i cross-sectional area of the ith element.

I_i moment of inertia of the ith element.

d^-, d^+ deviation variables.

Satisfy

System constraint

- Combined stress constraints

$S(A_i, I_i) \le S_{max}$, $i = 1,2,3$.

System goals

- Mass goal: minimize the mass of the system

$V(A_i)/V^E + d^- - d^+ = 1.0$, $i = 1,2,3$.

Bounds on parent system variables

$L^A \le A_i \le U^A$, $i = 1,2,3$.

$L^I \le I_i \le U^I$, $i = 1,2,3$.

Minimize

The Deviation Function (Archimedean form):

$Z = w_1 d^- + w_2 d^+$

Fig. 3.32 A cDSP formulation at parent system level [14]

marked as "Parent", which is linked to the three existing subsystem level ***Process*** Instances "M1", "M2", and "M3" by "V1", "V2" and "V3", respectively. Also, the lateral interactions, "V1", "V2", and "V3" represent strong coupling with two-way information flows. The specification of "V1" is shown in the window under the canvas of Fig. 3.33. Details of the information flows are embodied in Slots "Flow_1_to_2" and "Flow_2_to_1". Deviation variables related to the interaction system goals are incorporated in the deviation function in a preemptive form. The vertical interaction

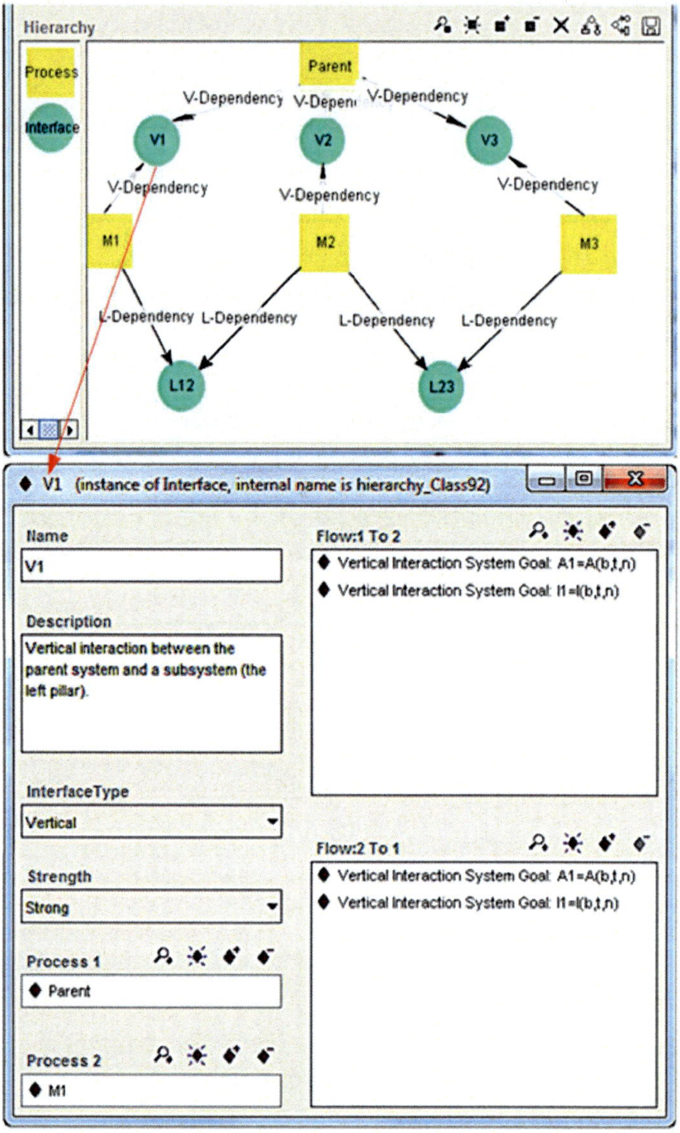

Fig. 3.33 Specification of the comprehensive model [14]

goals are of the highest priority, the lateral interaction goals are the second priority, and the mass goal is the third. Specification of the deviation function is assigned to the "root" node of the hierarchy, namely, "Parent" in Fig. 3.33, being the overall control for the model. At a computational level, all the hierarchy information is integrated as a coupled cDSP and computed with DSIDES. The computed results are shown in Table 3.19, highlighting that the subsystem dimensions match each other very well, similarly to Table 3.18. This is because lateral interactions are included in the

Table 3.19 Results of the comprehensive model [14]

Parent system (A-cm^2, I-cm^4)					
Variable	Value	Variable	Value	Variable	Value
A_1	40.21	A_2	40.18	A_3	40.16
I_1	20,445	I_2	20,419	I_3	20,426
Member 1 (cm)		Member 2 (cm)		Member 3 (cm)	
Variable	Value	Variable	Value	Variable	Value
b_1	12.51	b_1	12.5	b_1	12.49
b_2	13.1	b_2	13.1	b_2	13.12
t_1	0.555	t_1	0.554	t_1	0.554
t_2	0.583	t_2	0.583	t_2	0.583
t_3	0.434	t_3	0.433	t_3	0.433
h	59.1	h	59.09	h	59.08
Volume (cm^3): 100,480					

model. This has been verified by manually checking that the values of the parent system-level variables also match those of the subsystem level variables (e.g., cross-sectional value of A matches the value of $A(b, t, h)$). This illustrates the importance of including the vertical interactions in the model. The final volume of the portal frame is increased from 98,605 to 100,480 cm^3. The difference is partially due to the bounds (especially the lower bounds) assigned to the parent system-level variables and partially to the highest priority assigned to the vertical interaction goals.

3.5 Empirical Structural Validity

Empirical structural validation is to build confidence in the appropriateness of the test example problems chosen for illustrating and verifying the performance of the framework and methods. The ontology for representing selection decision-related knowledge is first tested using a light switch cover plate material selection example. In the example, different scenarios including the documentation of the original material selection, adding new material alternatives, adding new attributes for consideration, and parameter variation are considered. Using the ontology-based selection DSP template, the reusability, executability, and consistency of selection-related knowledge are tested. The example thus is appropriate to demonstrate the utility of the ontology for selection decisions as they involve multiple alternatives and attributes, as well as the variation (which is critical for testing knowledge reusability) of them in the selection process. Using the second example, the ontology for representing the knowledge related to compromise decisions is tested using a pressure vessel design problem. The reusability, executability, and consistency of the ontology are verified in both original design and variant design (including reduction of the dimensions of the design space, increase of the number of goals, and changes in the weights of goals) of the pressure vessel. Therefore, the example is suitable to demonstrate the utility of the ontology for compromise decisions because they involve multiple variables, constraints, and goals, as well as their variation in variant design. In the third example, we use the portal frame design problem to test the decomposability, flexibility, and visualization capability of the ontology for representing the knowledge related to coupled hierarchical decisions. The portal frame example involves two level of decisions that are coupled at the same level and cross levels. In the variant design scenarios, the lateral and vertical interactions are gradually added to the original all-in-one cDSP, and we see that the ontology is very flexible to support the decomposition of the original cDSP, the refinement of interaction information among sub cDSPs, and the visualization of the whole hierarchical cDSP structure. Based on the features of the problem and the performance of the ontology, it is concluded that the portal frame design example is appropriate.

3.6 Where We Are and What Comes Next?

In this chapter, we present three ontologies for representing the knowledge related to selection decisions, compromise decisions, and coupled hierarchical decisions, that serve as the foundation for providing knowledge support in individual decision-making on the PDSIDES platform. In the next chapter, we describe ontologies for representing knowledge related to decision workflows.

References

1. Gruber, T. R. (1993). A translation approach to portable ontology specifications. *Knowledge Acquisition, 5*(2), 199–220.
2. Lu, W. L., Qin, Y. C., Liu, X. J., Huang, M. F., Zhou, L. P., & Jiang, X. Q. (2015). Enriching the semantics of variational geometric constraint data with ontology. *Computer Aided Design, 63*, 72–85.
3. Barbau, R., Krima, S., Rachuri, S., Narayanan, A., Fiorentini, X., Foufou, S., & Sriram, R. D. (2012). OntoSTEP: Enriching product model data using ontologies. *Computer Aided Design, 44*(6), 575–590.
4. Li, Z., Raskin, V., & Ramani, K. (2008). Developing engineering ontology for information retrieval. *Journal of Computing and Information Science in Engineering, 8*(1), 011003.
5. Liu, Y., Lim, S. C. J., & Lee, W. B. (2013). Product family design through ontology-based faceted component analysis, selection, and optimization. *The Journal of Mechanical Design, 135*(8), 081007.
6. Witherell, P., Krishnamurty, S., & Grosse, I. R. (2007). Ontologies for supporting engineering design optimization. *Journal of Computing and Information Science in Engineering, 7*(2), 141–150.
7. Rockwell, J. A., Grosse, I. R., Krishnamurty, S., & Wileden, J. C. (2010). A semantic information model for capturing and communicating design decisions. *Journal of Computing and Information Science in Engineering, 10*(3), 031008.

8. Ming, Z., Wang, G., Yan, Y., Dal Santo, J., Allen, J. K., & Mistree, F. (2017). An ontology for reusable and executable decision templates. *Journal of Computing and Information Science in Engineering, 17*(3), 031008.
9. Keeney, R. L., & Raiffa, H. (1976). *Decisions with multiple objectives: Preferences and value tradeoffs*. Wiley.
10. Fernandez, M. G., Seepersad, C. C., Rosen, D. W., Allen, J. K., & Mistree, F. (2005). Decision support in concurrent engineering—The utility-based selection decision support problem. *Concurrent Engineering: Research and Applications, 13*(1), 13–27.
11. *Java Expert System Shell (JESS), the Rule Engine for the JavaTM Platform, 2015*. http://herzberg.ca.sandia.gov/
12. 2015, *Protégé 3.5 Release*. Stanford University. http://protegewiki.stanford.edu/wiki/Protege_3.5_Release_Notes
13. 2015. *Jess Tab*. Henrik Eriksson. http://www.jessrules.com/jesswiki/view?JessTab
14. Ming, Z., Yan, Y., Wang, G., Panchal, J. H., Goh, C. H., Allen, J. K., & Mistree, F. (2015). Ontology-based executable design decision template representation and reuse. *ASME IDETC/CIE Conference* Paper Number DETC2015–46272, Boston, MA.
15. Ming, Z., Yan, Y., Wang, G., Panchal, J. H., Goh, C.-H., Allen, J. K., & Mistree, F. (2016). Ontology-based executable design decision template representation and reuse. *AI EDAM—Artificial Intelligence for Engineering Design Analysis and Manufacturing, 30*(4), 390–405.
16. Panchal, J. H., Fernández, M. G., Paredis, C. J. J., & Mistree, F. (2004). Reusable design processes via modular, executable, decision-centric templates. In *AIAA/ISSMO Multidisciplinary Analysis and Optimization Conference*, Albany, NY. Paper Number AIAA-2004-4601.
17. Mistree, F., Hughes, O. F., & Bras, B. A. (1993). The compromise decision support problem and the adaptive linear programming algorithm. In M. P. Kamat (Ed.), *Structural optimization: Status and promise* (pp. 247–286). AIAA.
18. Lewis, K., & Mistree, F. (1998). Collaborative, sequential, and isolated decisions in design. *Journal of Mechanical Design, 120*(4), 643–652.
19. Sandgren, E. (1990). Nonlinear integer and discrete programming in mechanical design optimization. *The Journal of Mechanical Design, 112*(2), 223–229.
20. Kulkarni, N., Gautham, B., Zagade, P., Panchal, J., Allen, J. K., & Mistree, F. (2015). Exploring the geometry and material space in gear design. *Engineering Optimization, 47*(4), 561–577.
21. Shukla, R., Kulkarni, N., Gautham, B., Singh, A., Mistree, F., Allen, J., & Panchal, J. (2015). Design exploration of engineered materials, products, and associated manufacturing processes. *The Journal of The Minerals Metals & Materials Society (TMS), 67*(1), 94–107.
22. Kuppuraju, N., Ganesan, S., Mistree, F., & Sobieski, J. S. (1985). Hierarchical decision-making in system-design. *Engineering Optimization, 8*(3), 223–252.
23. Shupe, J. A., Mistree, F., & Sobieski, J. S. (1987). Compromise—An effective approach for the hierarchical design of structural systems. *Computers & Structures, 26*(6), 1027–1037.
24. Shupe, J. A., Allen, J. K., & Mistree, F. (1987). Compromise: An effective approach for the design of damage tolerant structures. *Computers and Structures, 27*(3), 407–415.
25. Bascaran, E., Bannerot, R., & Mistree, F. (1987). The conceptual development of a method for solving multi-objective hierarchical thermal design problems. In *ASME 1987 National Heat Transfer Conference*, Pittsburgh, Pennsylvania. Paper Number 87-HT-62.
26. Bascaran, E., Bannerot, R. B., & Mistree, F. (1989). Hierarchical selection decision support problems in conceptual design. *Engineering Optimization, 14*(3), 207–238.
27. Allen, J. K., Krishnamachari, R. S., Masetta, J., Pearce, D., Rigby, D., & Mistree, F. (1992). Fuzzy compromise—An effective way to solve hierarchical design-problems. *Structural Optimization, 4*(2), 115–120.
28. Vadde, S., Allen, J. K., & Mistree, F. (1994). The Bayesian compromise decision-support problem for multilevel design involving uncertainty. *The Journal of Mechanical Design, 116*(2), 388–395.
29. Sobieszczanski-Sobieski, J. (1982). *A linear decomposition method for large optimization problems. Blueprint for Development—NASA-TM-83248*. Sponsoring Organization: NASA Langley Research Center.

30. Ming, Z., Yan, Y., Wang, G., Panchal, J. H., Goh, C. H., Allen, J. K., & Mistree, F. (2016). Ontology-based executable design decision template representation and reuse. *Ai Edam Artificial Intelligence for Engineering Design Analysis & Manufacturing, 30*(4), 390–405.
31. Ming, Z., Wang, G., Yan, Y., Dal Santo, J., Allen, J. K., & Mistree, F. (2017). An ontology for reusable and executable decision templates. *Journal of Computing and Information Science in Engineering, 17*(3), 031008.
32. Karandikar, H. M. (1989). *Hierarchical decision making for the intergration of information from design and manufacturing processes in concurrent engineering*. Ph.D., University of Houston, Houston, TX.
33. 2016, *Graph Widget of Protégé*. Standford University. http://protegewiki.stanford.edu/wiki/Graph_Widget_Tutorial_OWL

Chapter 4
A Platform for Decision Support in the Design of Engineered Systems (PDSIDES) and Design of a Hot Rod Rolling System Using PDSIDES

In this chapter, we present a knowledge-based Platform for Decision Support in the Design of Engineered Systems (PDSIDES). PDSIDES is built on the ontologies developed in Chap. 3, which are the primary constructs of the platform for decision support. Based on these primary constructs, we introduce the overview of PDSIDES, users and working scenarios, and the knowledge-based decision support modes in PDSIDES. Then, we describe how PDSIDES is implemented to deliver functionalities. Furthermore, in this chapter we test PDSIDES performance via a gear manufacturing-process design problem, i.e., a complex system design requiring several decisions. From the raw material to the final gear product, the material goes through multiple unit operations such as casting, rolling, cooling, forging, and machining, which are some of the processes in the steel manufacturing chain. To obtain the desired end-properties of the gear produced, proper decisions need to be made about the process control parameters (*set points*) during each of these processes. Many plant trials involving time and cost are needed to identify these operating set points. An alternative to this is exploiting the advancements in modeling tools and frameworks to design the system and realize the end product. Decisions made at each manufacturing unit are formulated as cDSPs and linked as a decision network that is mathematically modeled as coupled cDSPs, using a goal-oriented, inverse decision-based design method. A hot rod rolling system design problem example is used to test the performance of PDSIDES. A summary of this chapter is presented in Table 4.1. The mapping of the sections to the components (topics) discussed in this chapter is presented in Row 2 of Table 4.1.

4.1 Primary Constructs of PDSIDES

In Chap. 2, we state that our implementation of Decision-Based Design (DBD) is the Decision Support Problem Technique (DSPT), wherein the selection Decision Support Problem (sDSP) and the compromise Decision Support Problem (cDSP)

© The Author(s), under exclusive license to Springer Nature Switzerland AG 2022
Z. Ming et al., *Architecting A Knowledge-Based Platform for Design Engineering 4.0*,
https://doi.org/10.1007/978-3-030-90521-7_4

Table 4.1 Summary interpretation of the problem investigated in Chap. 4

Elements	**What?**
Components	Prototype of PDSIDES, including platform design (Sects. 4.1–4.2), platform implementation (Sect. 4.3), and platform performance test (Sects. 4.4–4.9)
Connectors	Platform PDSIDES
Form	**How?**
Component roles	Provide user-specific knowledge support to different types of users
Properties	An integrated decision support platform
Relationship	The implementation of the constructs for supporting selection, compromise, and coupled decisions in design
Rationale	**Why?**
Motivation	Providing supports to different types of users based on their knowledge levels
Assumptions	There are three types of users: template creators, template editors, and template implementers
Interpretation	Understanding how different types of users are supported in their decision-based design activities using PDSIDES

are the two primary constructs for formulating decisions in design. Any complex design can be represented through modeling a network of compromise and selection decisions. Our goal in designing PDSIDES is to facilitate designers with rapidly creating decision models for the specific design problems using the DSP constructs, making decisions, and finally, the produced decision knowledge can be stored and reused by other users for similar designs. To achieve this goal, the DSPs are represented as computational decision templates in PDSIDES. The templates include the sDSP template, cDSP template, and the coupled DSP templates, and their associated modules are managed in a module repository, as illustrated in Fig. 4.1. It is noted that the sDSP template and the cDSP template are also defined as a particular type of module since they comprise the key "building blocks" of a decision hierarchy and can be linked together using the *interface* and *process* modules (see Chap. 3 for details). Template modules represent the declarative knowledge in PDSIDES, which embodies problem-specific information and can be reused in the instantiation of DSP templates to support a designer making selection, compromise, and assist in making hierarchical coupled decisions. The procedural knowledge denotes how specific information is processed to reach a decision and is archived in the templates (the printed "wiring" between different modules) to execute decisions. The separation of these two types of knowledge makes it reasonably easy for designers to reconfigure existing templates, which is critically essential in adaptive and variant designs, where design consideration changes and the original decision model needs to be modified. Template modification is discussed in Sect. 4.2.

The definition of the templates and their modules are discussed in Chap. 3 using frame-based ontologies, which provide the foundation for representing the templates at a computational level. The advantages of employing ontology in PDSIDES are summarized as follows.

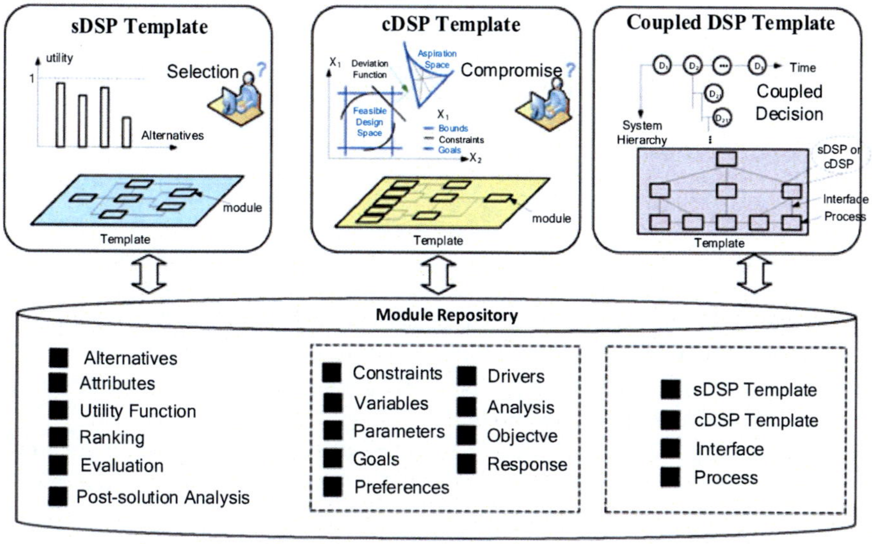

Fig. 4.1 DSP templates and their associated modules

- ***Facilitate knowledge sharing***. This is embodied in two aspects: knowledge sharing among different users in PDSIDES and knowledge sharing between PDSIDES and other Product Lifecycle Management (PLM) platforms. The DSP ontologies represent the common language used for design decision-making in PDSIDES. Thus, users from different design disciplines (e.g., thermal, structural, dynamic, etc.) can easily understand and communicate knowledge such as *variables, goals,* and *constraints*, etc., with each other. Meanwhile, the explicit, formal specifications of the terms in DSP ontologies enable PDSIDES to exchange knowledge with other PLM platforms such as product data management systems and simulation-based analysis systems.
- ***Facilitate knowledge population***. In order for the DSP templates to execute and effect decisions, the modules of the templates must be populated with specific knowledge or information. The DSP ontologies are the abstract representations of the templates, which is very convenient for instantiating different instances with specific information.
- ***Facilitate knowledge retrieval***. One of the prerequisites for the reuse of templates and the associated modules is that they can be retrieved from the repository, i.e., knowledge base when needed. The DSP ontologies capture the complex semantic relationships among the modules and templates, which allows them to support semantic-based retrieval that can respond to comprehensive query needs. For details regarding semantic retrieval, the reader is referred to [1].
- ***Facilitating consistency maintaining***. Modification of the original templates usually happens in adaptive or variant design, which may lead to inconsistency of the modified templates since the arrangement or values of the modules are

changed. The DSP ontologies support rule-based reasoning and appropriately handle the inconsistency.

4.2 Design of Platform PDSIDES

4.2.1 Platform Overview

An overview of PDSIDES is illustrated in Fig. 4.2. PDSIDES is divided into three parts: knowledge, users, and decision-based design. What follows is the description of the platform from a bottom-up perspective that includes how these three parts are connected to enable the functionalities.

At the bottom of PDSIDES, decision-related knowledge is stored in the knowledge base (including the module repository). The knowledge, including declarative

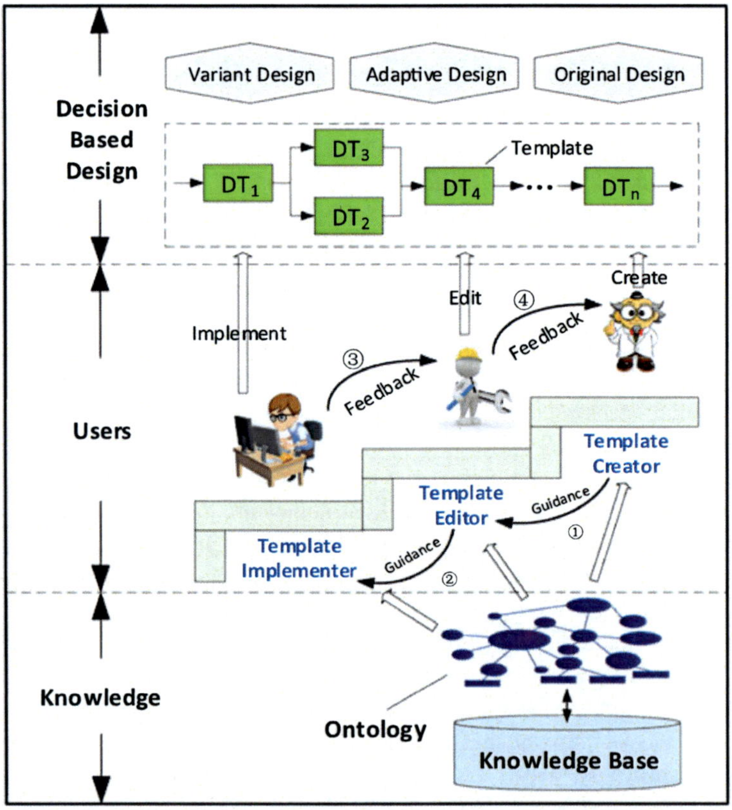

Fig. 4.2 PDSIDES overview [2]

knowledge such as *problem statement, alternatives, attributes, variables, parameters,* and *constraints*, etc., and procedural knowledge such as consistency rules and computing codes which are needed to calculate, e.g., the expected utility of an sDSP template, are organized by a holistic ontology which is the combination of the three ontologies developed in Chap. 3. The middle part contains the types of users, namely, the Template Creator, Template Editor, and Template Implementer, formally defined in Sect. 4.2.2. These three user types embody three different levels of knowledge, represented by the stairs in Fig. 4.2. The top level is the Template Creator responsible for creating the DSP templates, the middle level is the Template Editor responsible for editing DSP templates, and the bottom level is the Template Implementer responsible for implementing the DSP templates. The interactions among the three user types are a closed loop, where the template operational guidance is passed downwards from the Creator to the Editor, then to the Implementer, and the feedback of operating the templates is sent upwards from the Implementer to the Editor and then to the Creator. The creation, editing, and implementation of the DSP templates are all facilitated using a holistic ontology. The top part of PDSIDES is about decision-based design. In PDSIDES, design is classified into three types, namely, original design, adaptive design, and variant design; all are realized from a decision-based perspective using the DSP templates. In specific design cases, the underlying decision workflow is represented by networked DSP templates that can be exercised by the three types of users through creating, editing, and implementing.

4.2.2 Users and Working Scenarios

The definitions of the three user types are introduced, and their associated working scenarios are described in detail in this section.

Template Creator. Template Creators are domain experts responsible for creating DSP templates for original designs that call for new concepts. An original design usually needs the working principle of the system to be determined. In PDSIDES, to create an original design template, the Creators first need to determine the decision type that needs to be made since different types of decisions require different knowledge. For selection decisions, Creators need to define the selection alternatives, attributes to evaluate the alternatives, and utility functions to measure the performance of the alternatives. For compromise decisions, Creators need to identify the variables representing the system's features, constraints, and bounds that confine the feasible design space and goals and preferences that determine the aspiration space. For hierarchical coupled decisions, in addition to the determination of "nodes" in the decision workflow, which may be selection or compromise, Creators also need to identify the dependency and the associated information flows between different "nodes". The knowledge can be the Creators' previous experience, prediction, or results from simulation analysis. With this knowledge, template modules are created and assembled to form decision templates that are then tested and stored for reuse.

Template Editor. Template Editors are senior designers with sufficient knowledge and experience in a specific domain and are responsible for editing or tailoring existing decision templates in adaptive designs. This requires the original templates to be adapted for new applications. The adaptive design stands for cases where the system's working principle remains the same, while some design considerations vary due to the requirements' evolution. For example, a pressure vessel may need to be redesigned to adapt to a new goal of minimizing the economic cost because of intensive market competition. In PDSIDES, to perform an adaptive design, the Template Editors need to modify existing DSP templates to reflect the change of the design requirements. For the sDSP templates, the modification includes adding/removing *alternatives* and *attributes* and reconfiguring the *utility functions*. For the cDSP templates, the modification includes adding/removing *variables, constraints,* and *goals*. For the hierarchical coupled DSP templates, the modification includes three aspects: modifying the modules within the DSP templates in a decision workflow, modifying the number of DSP templates (adding/removing sDSP or cDSP templates), and modifying the arrangement (sequence, information flow, etc.) of the DSP templates. The Editor's knowledge related to the modification is captured in the newly modified DSP templates that are stored and used for new applications.

Template Implementer. Template Implementers are designers who have basic knowledge and typically little knowledge or interest in the analysis embodied in the template. They are responsible for executing existing decision templates that result in variant designs that require only parametric changes to the original decision templates. Variant designs usually happen when the values of some original design parameters alter. For example, assuming that some new materials replace the original material of a pressure vessel with different *density* and *strength*, the values of parameters *density* and *strength* of the original design model (e.g., the cDSP) need to be updated to reflect the change that will result in a different dimension of the pressure vessel. In PDSIDES, variant design Template Implementers can change the DSP template parameters values of: (1) bounds of the sDSP *attributes* or cDSP *variables*, (2) cDSP parameters and targets, and (3) the relative importance of the sDSP *attributes* and cDSP *goals*. By changing the parameters' values, Template Implementers can execute the DSP templates and obtain variant designs.

In PDSIDES, users with access to higher knowledge levels also can perform the operations defined for users of lower knowledge levels. For example, a Template Creator can be an Editor or Implementer, while an Editor can also be an Implementer. With decisions modeled as DSP templates and users classified into three types, the process of decision-based design in PDSIDES is presented in Fig. 4.3. A user, e.g., a domain expert, first describes the design problem and then searches PDSIDES for a DSP template to support the design. In PDSIDES, DSP template searching is a query-based process where a problem statement (a short text) is used as the input, and a documented DSP template instance is the output. The problem statement and template instances are mathematically represented using the bag-of-word approach [3] during the query process. The similarity between the problem statement and different template instances is measured by a cosine coefficient [4] as follows:

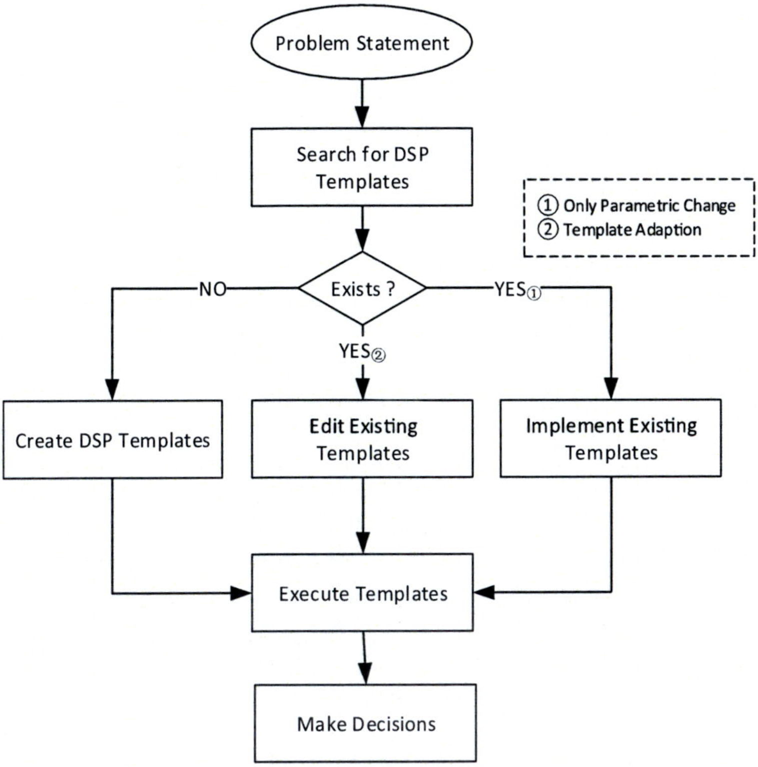

Fig. 4.3 Flowchart of decision-based design In PDSIDES [2]

$$sim(\grave{A}, \grave{B}) = \frac{\vec{A} \times \vec{B}}{\left|\vec{A}\right| \cdot \left|\vec{B}\right|}$$

where $\vec{A}$ and $\vec{B}$ are two n-dimensional vectors representing the word frequencies for the given problem statement and a specific template instance. It should be noted that the *bag-of-word* characterizing the template instance includes not only words from the textual slots such as "name" and "description", but also words from the structural slots such as "variables" and "constraints", which makes the instance more comprehensive and more easily matched. If no DSP template instance is matched, a new template needs to be created, executed, and exploited to make the final decision. If there is an existing template(s), the designer needs to determine how much modification needs to be made to the template. If changes involve only the value of a parameter, then the designer resets the parameter values, executes the template, and decides. If more adaption is needed, the designer needs to do the editing before executing the template and then deciding.

4.2.3 Knowledge-Based Decision Support Modes

The core of PDSIDES is the ontology that integrates the knowledge to support the three designer types, namely, Template Creators, Template Editors, and Template Implementers. Figure 4.4 represents how knowledge-based decision support is provided to the three designer types in their associated working scenarios (by taking the cDSP templates as an example).

Template Creators—*provide the vocabulary for modeling decisions and capture associated knowledge*. Template Creators need a formal language to help them describe and model the decisions for the original design. The DSP ontologies in PDSIDES can provide them with the vocabulary to model their decisions. For example, the term *variable* is defined as a Class with several slots, including *upper bound, lower bound, unit,* and *value*, which will help specify the module "*variables*" of the cDSP template. Using the classes and slots defined in the ontology, DSP templates can be quickly instantiated as instances, which are captured and stored in the database for reuse, as shown in the top-left picture of Fig. 4.4.

Template Editors—*ensure consistency for editing*. As mentioned earlier, modifying existing DSP templates may incur inconsistency, especially when the template is highly complex, e.g., tens of *variables, constraints,* or *goals*, and the Editor who modifies the template is not the original Creator and does not have complete knowledge about the template. Therefore, a consistency checking mechanism is required

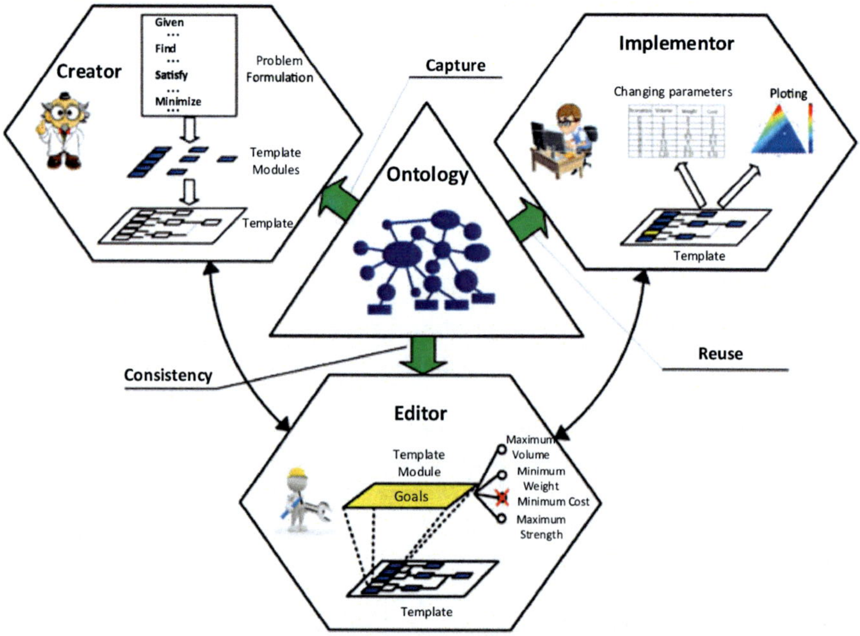

Fig. 4.4 Knowledge-based decision support in PDSIDES [2]

to identify the potential inconsistency. A rule-based reasoning mechanism is associated with the DSP ontologies in PDSIDES to provide a consistency checking service for the Template Editors. Rules are extracted from the DSP constructs, such as the sum of the weights assigned to the *goals* must equal one. An example of a Template Editor removing an existing goal (minimum cost) from the cDSP template is illustrated at the bottom of Fig. 4.4, where PDSIDES will check this inconsistency and inform the user accordingly.

Template Implementers—*reuse of the documented knowledge and perform post-solution analysis.* As stated in Sect. 4.2.2, Template Implementers have little knowledge or interest in the analysis embodied in the templates, and what they need is information that helps them exercise the template and make the decision. In PDSIDES, the knowledge provided to the Template Implementers includes both declarative and procedural knowledge. The former is captured from Template Creators and Editors, and the latter is built into the platform, for example design space exploration algorithms, plotting routines, etc., which are hard-coded and can be invoked when needed. The picture on the top-right in Fig. 4.4 represents a Template Implementer changing the weights assigned to different goals and using the ternary plot to identify the insensitive weight sets to make a robust decision and the knowledge documented in the template is reused.

4.3 Implementation of Platform PDSIDES

PDSIDES is implemented as a two-tier client–server architecture to provide knowledge-based decision support with web browser-based graphical user interfaces (GUI) over the internet, as shown in Fig. 4.5. In the client–server architecture, applications of PDSIDES are deployed to a web application server (marked as "Knowledge Server" in Fig. 4.5) and provide remote user access using browsers such as Internet Explorer or Google Chrome. Due to the easy access through web browsers, PDSIDES is affordable for many users involved in the decision template creation, editing, and executing the process for an engineering system design. The entire design process becomes a knowledge capturing, evolution, and reuse process over the internet. Maintenance and upgrades for PDSIDES in a client–server architecture are convenient since the application package is deployed in one web server instead of being distributed to a wide range of client computers. The client-side of PDSIDES is the user interaction GUI, including a template for searching and browsing that is designed for locating the desired DSP templates and presenting them, a template for creating and editing that is designed to instantiate and modify the DSP templates, and a template execution and analysis GUI, designed for executing DSP templates and performing post-solution analysis. The GUI can communicate with the PDSIDES Knowledge Server by a request-response mode using the HyperText Transfer Protocol (HTTP). PDSIDES Knowledge Server includes five main parts: Response Server, Knowledge Base, JESS Reasoner, DSIDES, and Matlab. The Response Server is the central "brain" that integrates the other four parts responding to requests. The Response Server itself

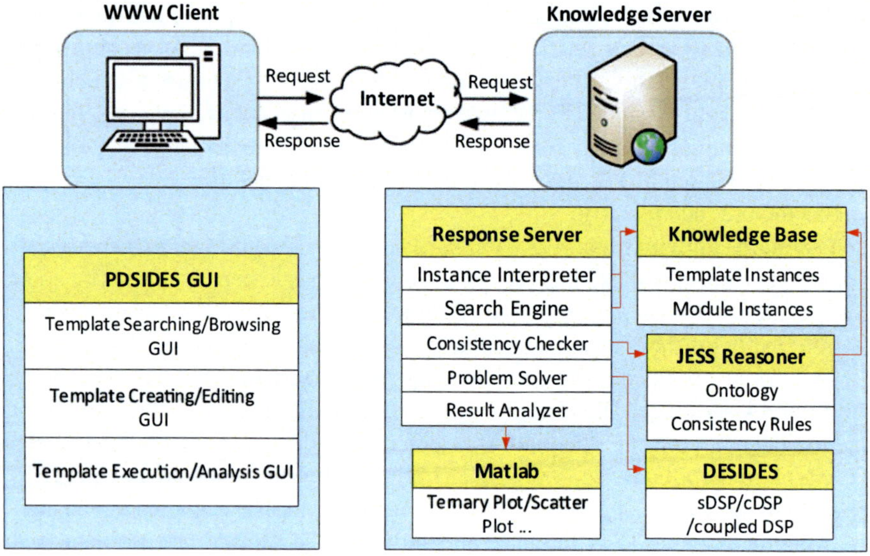

Fig. 4.5 System Architecture of PDSIDES [2]

has five components: a search engine, an instance Interpreter, a consistency checker, problem solver, and a result analyzer.

The instance interpreter interprets the data collected from the Template Creators (or Editors) and formats them into DSP Template instances according to the DSP ontologies. The generated template instances and module instances are stored in the Knowledge Base. The search engine is connected to the Knowledge Base to provide ontological semantic-based knowledge retrieval. Consistency checking is facilitated through a consistency checker and the JESS Reasoner—the Rule Engine for the JavaTM Platform [5], which can provide rule-based intelligence inference. The problem solver is connected to DSIDES for solving the DSPs and is invoked when a template executer executes a template. The result analyzer is to help users, especially Template Implementers, to analyze the results produced by the problem solver. Finally, Matlab has a strong capability in providing data visualization tools such as ternary plots and scatter plots, and therefore this feature is integrated into PDSIDES.

The front-end, i.e., GUI, of PDSIDES is realized by JavaScript embedded in web pages. The development process is facilitated by Sencha Inc.'s GXT [6]. GXT is a comprehensive Java framework that uses the GWT (Google Web Toolkit) compiler [7], allowing developers to write applications in Java and compile their codes into highly optimized JavaScript that supports feature-rich web applications. To enable graph-based interaction in terms of the operation of the DSP networks that may have multiple DSP templates and associated connections involved, Apache Flex [8], a rich internet application developing framework, is integrated into GXT to facilitate the creation of web-based diagrams. A DSP template such as a cDSP template may be very complex and have tens of *variables, parameters, constraints,* and *goals*, which often overloads data transmission between the front-end and the back-end. To address this issue, JSON [9], a lightweight data-interchange format, is used as the data transmission scheme together with the HTTP protocol.

The back-end, i.e., the server-side, of PDSIDES is written in Java to enable interoperability among different applications and cross-platform deployment. Many back-end applications such as the instance interpreter, search engine, consistency checker, and JESS reasoner heavily depend on the DSP ontologies. As discussed in Chap. 3, the DSP ontologies are formalized using the frame-based paradigm that contains Classes and Slots. The realization of this paradigm using the frame language is presented in Fig. 4.6. The top box in the figure represents Class "*SystemGoal*" in the cDSP ontology, which includes definitions of slots such as *target, linearity,* and *equality*, and the associated facets include *type, cardinality,* and *allowed-values*. The frame-based ontology is actually an object-oriented mechanism based on which lots of instances can be populated. Two boxes at the bottom of Fig. 4.6 represent two instances, i.e., *volume goal* and *weight goal*, of the Class "*SystemGoal*" represented using frame language. The specific data in the slots of the instances are first collected using the template creating/editing GUI, then processed by the instance interpreter, and finally retained in relational databases (in PDSIDES, we use Oracle). Instances are treated as facts that are processed in the consistency checking process. In the JESS reasoner, all the facts are matched to the consistency rules and take specific actions if the corresponding rules are triggered. An example of the consistency rules is as follows:

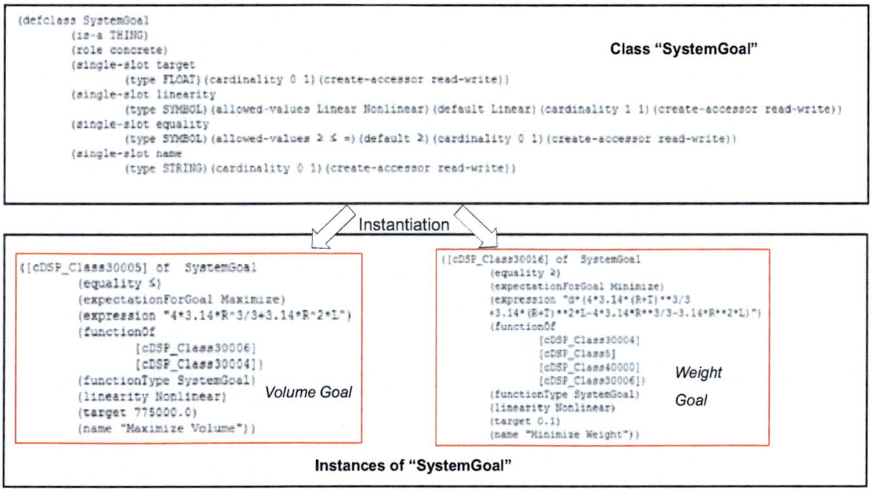

Fig. 4.6 Frame-based realization of the ontology and associate instances [2]

Fig. 4.7 PDSIDES portal [2]

```
(defrule MAIN::rule_5.1
  (object (is-a cDSPTemplate) (OBJECT ?a))
  =>
  (bind ?k (Sum (slot-get ?a preference) ONE))
  (if (and (<> ?k 1.0) (<> ?k 0.0)) then (printout t "MESSAGE: the sum of all the preferences is not 1.0!"
crlf)))
```

This rule means that if the sum of all preferences, i.e., weights, in any instance of Class cDSPTemplate is not equal to one, the reasoner will send a message about this inconsistency to the user operating the template instance.

The portal of PDSIDES is illustrated in Fig. 4.7. A user can log in to PDSIDES through a web browser using a username and password. Template Creators, Editors, and Implementers are three roles assigned to the users according to their knowledge in a specific domain, while a designer can have more than one role. Each role has its particular platform setup, and the portal is the view shared by all three roles. The portal includes two main parts, the left-hand side is the navigation panel, and the right-hand side is the statistical information panel. The former represents the critical functionalities of PDSIDES, including the Decision Knowledge Management portion (managing knowledge about selection, compromise, and hierarchical decisions. Access is assigned to Creators and Editors), the Design Decision Support portion (providing the DSP template execution and analysis service, assigned to Implementers), and the settings portion (purview management, access is assigned only to system administrators). The latter presents the charts and tables in terms of the decision-related knowledge and users. Users can see the number and the distribution of DSP templates in PDSIDES, the ranking of Creators who contribute their knowledge to PDSIDES, the ranking of templates that are reused frequently, and the latest updated DSP templates. They can also search, browse, and execute specific templates.

4.4 Testing the Performance of PDSIDES—Hot Rod Rolling Example Problem

The performance of platform PDSIDES is tested utilizing a gear manufacturing-process design problem, i.e., a complex system design that calls for a series of decisions. The foundational problem is contributed by our industrial partner, the Tata Consultancy Services in India [10]. From the raw material to the final gear product, the material goes through multiple unit operations such as casting, rolling, cooling, forging, and machining, which are some of the steel manufacturing-process chain processes. To obtain the desired end properties of the produced gear, some proper decisions must be made about the process control parameters (*set points*) during each process, with a large number of plant trials involving time and cost are required to identify these operating set points. An alternative solution is exploiting the advancements in modeling tools and frameworks to carry out the system's design to realize the end product. To couple the material processing-structure–property-performance spaces, we integrate the vertical and horizontal models, which further make it possible to carry out the integrated decision-based design of the manufacturing processes to realize the end product [11–17]. Decisions to be made at each manufacturing unit are formulated as cDSPs and are linked as a decision network (mathematically modeled as coupled cDSPs) using a goal-oriented, inverse decision-based design method [11]. In this chapter, the hot rod rolling system-design problem by Nellippallil et al. [11, 17, 18] is modified and exploited as an example to demonstrate the performance of PDSIDES. As mentioned earlier, the problem includes multiple stages. For the sake of simplicity, we frame a boundary within the cooling stage and the final rod product requirements.

4.5 Hot Rod Rolling System (HRRS) Design Problem

One of the critical requirements in a gear manufacturing process chain is minimizing the dishing/distortion during the gear blanks' forging stage. A significant factor influencing the distortion in forged gear blanks is the banded microstructure that occurs due to the creation of micro segregates (caused by the segregation of alloying elements like manganese) and processing carried out at the rolling and cooling stages. Managing the processing and the microstructural factors associated with banding will indirectly affect the final mechanical properties of the product. Hence, there is a need to predict the final mechanical properties of the product (rod) as a function of composition, microstructure, and the rolling and cooling parameters, satisfying the requirement of managing the banded microstructure. To predict the mechanical properties, several model integrations need to be implemented (vertical across scales and horizontal across processes), starting from rolling and cooling up to the end rod product, as illustrated in Fig. 4.8.

Nellippallil et al. [11] demonstrate applying a goal-oriented, inverse, decision-based design method using the cDSP construct. They integrate the vertical and horizontal models for the gear manufacturing process chain's hot rolling and cooling stages to produce a rod with defined properties. The mechanical property goals and requirements for yield strength (YS), tensile strength (TS), toughness measured by impact transition temperature (ITT), and hardness (HV) are identified for the rod. These mechanical properties are dependent on the final microstructure after cooling, like the ferrite grain size after cooling (FGS, D_α), the phase fractions of ferrite (X_f) and pearlite ($1-X_f$) the pearlite interlamellar spacing (S_0) and the composition variables like silicon (Si), nitrogen (N), phosphorous (P), and manganese (Mn). Again, these microstructure factors after cooling are defined by the cooling rate (CR) at

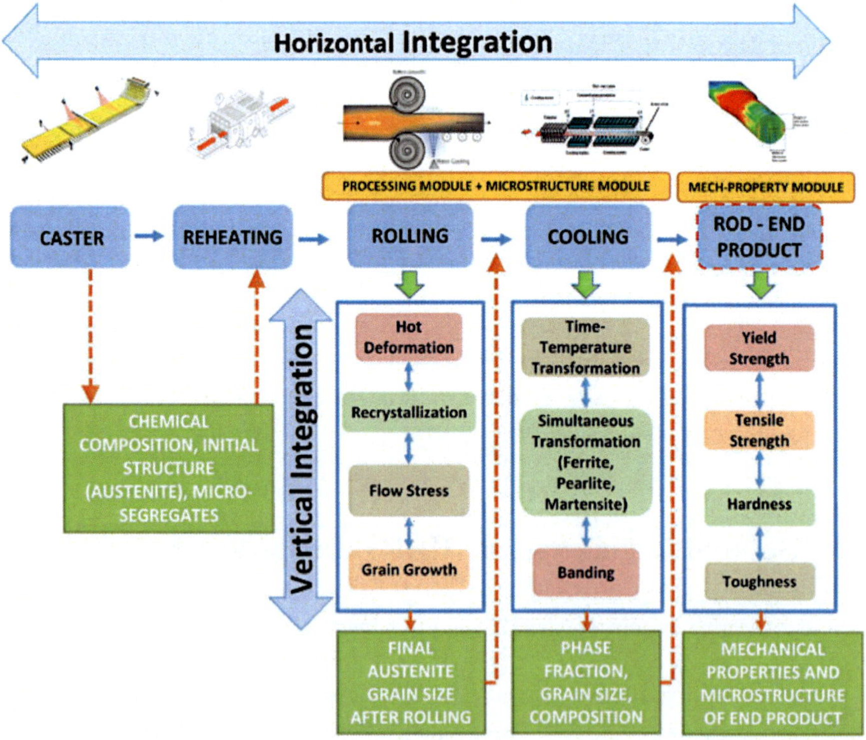

Fig. 4.8 Vertical and horizontal integration of models and information flow for hot rod rolling problem [14, 17]

which cooling is carried out, the final austenite grain size after rolling (AGS), and the composition variables like carbon (C) and manganese (Mn). The AGS is defined by the processing carried out at the rolling stage which requires the modeling of hot deformation, recrystallization, grain growth, etc., for its prediction. In this study, we frame a boundary within the cooling stage and the rod end product requirements. The inputs to the cooling stage are the AGS, CR, C, and Mn from the rolling process, and the outputs are $\boldsymbol{D_\alpha}$, $\boldsymbol{X_f}$, and $\boldsymbol{S_0}$, which, along with the composition variables, define the YS, TS, ITT, and HV of end rod produced. The models used to formulate the hot rod rolling problem are summarized in Table 4.2.

4.6 Knowledge-Based Decision Support in the Design of HRRS

On the platform, the process designers are divided into three groups [2], namely (1) ***Template Creators:*** They create single cDSP templates corresponding to each operation of the gear manufacturing process using the terminologies defined in the ontology and link these cDSP templates as decision network templates. (2) ***Template Editors:*** They edit the existing templates when the requirements of the final gear product are changed, e.g., new requirements are added. (3) ***Template Implementers:*** They execute existing templates by specifying parameter values. Here, we test the performance of PDSIDES in terms of template creation, consistency checking, and post-solution analysis.

4.7 Original Design

In the original design, the template creator (domain expert) formulates in PDSIDES the cDSP for the problem boundary framed within the hot rod rolling process chain problem by considering the complete information flow across models, thereby establishing relationships. The relationships established in the original design cDSP are the end mechanical properties of the product; *YS* (yield strength), *TS* (tensile strength), *ITT* (impact transition temperature) and *HV* (hardness) as a function of the system variables that are the output after rolling and are input to the cooling stage. The output parameters after cooling are FGS (ferrite grain size, D_α, X_f (phase fractions of ferrite), S_0 (pearlite interlamellar spacing). Moreover, composition variables that define the end mechanical properties are defined as constraints in the formulated cDSP. The end product mechanical property goals, e.g., maximizing *YS*, *TS* and minimizing *ITT*, along with the goal for managing banding by maximizing ferrite fraction, are captured in the cDSP. These goals are controlled by the independent system variables of this problem, namely *CR* (cooling rate), *AGS* (grain size after rolling), *C* (carbon), and *Mn* (manganese). The upper and lower limits for the system variables

Table 4.2 Models used to formulate hot rod rolling problem

Quantity	Model	References
Yield strength	$YS = X_f\left(77.7 + 59.9 \times [Mn] + 9.1 \times (0.001D_\alpha)^{-0.5}\right) + 478[N]^{0.5} + 1200[P] + (1 - X_f)[145.5 + 3.5S_0^{-0.5}]$ where YS is in MPa, S_o in µm, D_α in µm	Kuziak and co-authors [19]
Tensile strength	$TS = X_f\left(20 + 2440 \times [N]^{0.5} + 18.5 \times (0.001D_\alpha)^{-0.5}\right) + 750(1 - X_f) + 3\left(1 - X_f^{0.5}\right)S_0^{-0.5} + 92.5 \times [Si]$ where TS is in MPa, S_o in µm, D_α in µm	Kuziak and co-authors [19]
Hardness	$HV = X_f(361 - 0.357T_{mf} + 50[Si]) + 175(1 - X_f)$ Average austenite to ferrite transformation temperature (T_{mf}) is assumed as 700 °C	Yada [20]
Impact transition temperature	$ITT = X_f\left(-46 - 11.5D_\alpha^{-0.5}\right) + (1 - X_f)(-335 + 5.6S_0^{-0.5} - 13.3p^{-0.5} + \left(3.48 \times 10^6\right)t) + 49[Si] + 762[N]^{0.5}$ where D_α, S_o, p and t are in mm. We have assumed the value of pearlite colony size p as 6 µm and carbide thickness t as 0.025 µm	Gladman and co-authors [21]
Allotriomorphic ferrite	$X_{fa} = 1.59 - 0.26[C] - 0.00856CR - 0.0105D - 3.08[C] + 0.000826[Mn]CR + 0.0009[Mn]D + 0.7647[Mn][C] + 0.000011CR * D + 0.002CR[C] + 0.0032D[C] - 0.05058[Mn]^2 + 0.00004CR^2 + 0.000036D^2 + 2.483[C]^2$	Nellippallill and co-authors [14, 17]

(continued)

and the maximum and minimum values for specific cooling stage parameters, and the mechanical properties are defined in the cDSP as bounds and constraints. The target values for the goals are defined as YS_{Target}= 400 MPa, TS_{Target}= 780 MPa, ITT_{Target}= −90 °C, $X_{f\,Target}$= 0.8. The original design cDSP reads as follows:

Table 4.2 (continued)

Pearlite	$X_p = 0.206 - 0.117[Mn] - 0.0005CR - 0.00113D + 0.248[C] + 0.00032[Mn]CR + 0.000086[Mn]D + 0.9539[Mn][C] - 4.259 \times 10^{-6}CR * D + 0.00726CR[C] + 0.0023D[C] - 0.0305[Mn]^2 - 0.0000056CR^2 + 4.859 \times 10^{-6}D^2 + 0.79[C]^2$	Nellippallill and co-authors [14, 17]
Widmanstätten ferrite	$X_{fw} = 1 - (X_{fa} + X_p)$	Nellippallill and co-authors [14, 17]
Total ferrite	$X_f = (X_{fa} + X_{fw})$	Nellippallill and co-authors [14, 17]
Ferrite grain size	$D_\alpha = \left(1 - 0.45\varepsilon_r^{0.5}\right) \times \{(-0.4 + 6.37C_{eq}) + (24.2 - 59C_{eq})CR^{-0.5} + 22[1 - \exp(-0.015D)]\}$ for $C_{eq} < 0.35$ $D_\alpha = \left(1 - 0.45\varepsilon_r^{0.5}\right) \times \{(22.6 - 57C_{eq}) + 3CR^{-0.5} + 22[1 - \exp(-0.015D)]$ for $C_{eq} > 0.35$ where D is the final austenite grain size after rolling and ε_r is retained strain. C_{eq} is the carbon equivalent defined as $C_{eq} = (C + Mn)/6$	Hodgson and Gibbs [22]
Pearlite interlamellar spacing	$S_o = 0.1307 + 1.027[C] - 1.993[C]^2 - 0.1108[Mn] + 0.0305CR^{-0.52}$	Kuziak and co-authors [19]

Given

(1) End requirements identified for the rod rolling process

- Maximize Yield Strength (Goal)
- Maximize Tensile Strength (Goal)
- Minimize ITT (Goal)
- Maximize Ferrite Fraction (Goal)
- Maximize Hardness (Requirement)

Table 4.3 System variables and ranges for cDSP [2]

Sr. No.	System variables	Ranges
1	X_1, Cooling rate (CR)	11–100 K/min
2	X_2, Austenite grain size (AGS)	30–100 μm
3	X_3, the carbon concentration ($[C]$)	0.18–0.3%
4	X_4, the manganese concentration ($[Mn]$)	0.7–1.5%

(2) Well-established empirical and theoretical correlations, RSMs, and complete information flow from the end of rolling to the end product mechanical properties (more description in Refs. [14, 17])
(3) System variables and their ranges for the original design cDSP (Table 4.3)

Find
System Variables
X_1, Cooling Rate (CR)
X_2, Austenite Grain Size (AGS)
X_3, the carbon concentration ($[C]$)
X_4, the manganese concentration ($[Mn]$)
Deviation Variables
$d_i^-, d_i^+, i = 1,2,3,4$
Satisfy
System Constraints

- Minimum ferrite grain size constraint
- Maximum ferrite grain size constraint
- Minimum pearlite interlamellar spacing constraint
- Maximum interlamellar spacing constraint
- Minimum ferrite phase fraction constraint (manage banding)
- Maximum ferrite phase fraction constraint (manage banding)
- Minimum manganese concentration constraint (manage banding)
- Maximum manganese concentration constraint (manage banding)
- Maximum carbon equivalent constraint (manage banding)
- Minimum yield strength constraint
- Maximum yield strength constraint
- Minimum tensile strength constraint
- Maximum tensile strength constraint
- Minimum hardness constraint
- Maximum hardness constraint
- Minimum ITT constraint
- Maximum ITT constraint

System Goals
Goal 1:

- Maximize Yield Strength

$$\frac{YS(X_i)}{YS_{Target}} + d_1^- - d_1^+ = 1$$

Goal 2:

- Maximize Tensile Strength

$$\frac{TS(X_i)}{TS_{Target}} + d_2^- - d_2^+ = 1$$

Goal 3:

- Minimize ITT

$$\frac{ITT_{Target}}{ITT(X_i)} - d_3^- + d_3^+ = 1$$

Goal 4:

- Maximize Ferrite Fraction

$$\frac{X_f(X_i)}{X_{f\,Target}} + d_4^- - d_4^+ = 1$$

Variable Bounds
Defined in Table 4.3
Bounds on deviation variables

$$d_i^-, d_i^+ \geq 0 \text{ and } d_i^- * d_i^+ = 0, \text{ i} = 1,2,3$$

Minimize
We minimize the deviation function

$$Z = \sum_{i=1}^{4} W_i\left(d_i^- + d_i^+\right); \sum_{i=1}^{4} W_i = 1$$

By utilizing the cDSP formulation, knowledge associated with the following inverse the problem statement becomes: *Given the end product mechanical properties of a new steel product mix, what should be the microstructure after rolling and the design set points for the cooling stage that satisfies the requirements identified?* To facilitate the knowledge capturing process in the computational environment, PDSIDES provides a GUI for the template creator to create the required DSP templates (Fig. 4.9).

The building blocks on the left-hand side of the canvas, including Process and Interface, are formally defined in the ontology to create decision network templates (Hierarchical DSP templates). Since there is only one cDSP formulated for the original design of HRRS, the template creator can instantiate a *process* on the canvas and embody it with a cDSP template. The cDSP template is created in the "Compromise Decision Template Base" portion of PDSIDES. As shown in the window on the top-right of Fig. 4.9, the template creator can instantiate the HRRS cDSP template by specifying the slots including *name, problem statement, variables, parameters, constraints, goals,* and *preferences* using data such as Cooling Rate, Austenite Grain Size, and Carbon Concentration of the HRRS cDSP. As shown in the two panels at the GUI's bottom, facet information of the slots, such as *symbol, unit* of a *variable,* and *equation,* the *limit* of a *constraint*, are further specified using the GUI designed for the instantiation of template modules. When the HRRS cDSP template is populated with specific information, it is sent to the knowledge server for consistency checking, calculating the results, retention in the knowledge base, and availability for future reuse in adaptive and variant designs.

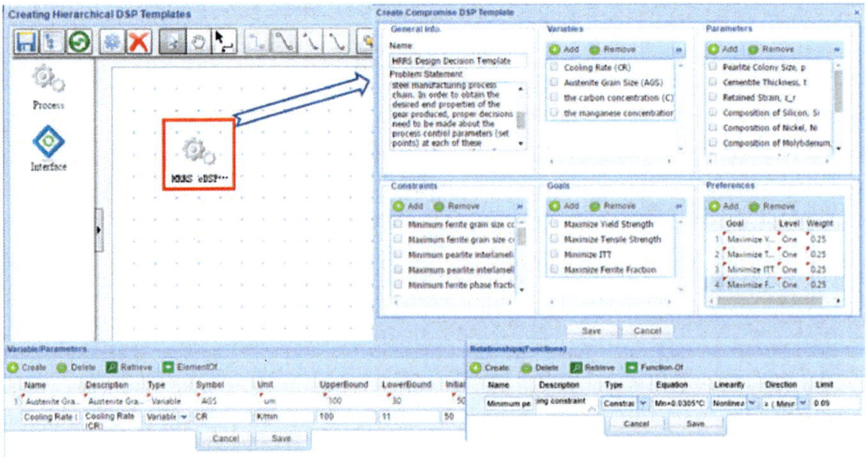

Fig. 4.9 Creating the HRRS design decision template in PDSIDES [2]

4.8 Adaptive Design

The template editor (senior designer) modifies the existing original design cDSP template according to the new requirements in an adaptive design. In the hot rod rolling problem addressed, the cDSP template of the original design relates the end product mechanical properties as a function of the microstructure factors after rolling and the cooling stage operating parameters. The intermediate factors, e.g., the ferrite grain size after cooling and the pearlite interlamellar spacing, which directly influence mechanical properties, are defined as constraints. Suppose the designer is interested in knowing the range of microstructure factors after cooling that will satisfy a given end mechanical property requirement. In this case, a new decision model must be created by considering the microstructure factors after cooling as independent variables to define the end mechanical properties. This requirement can be easily satisfied by editing the existing formulated original design cDSP template in PDSIDES. The editing involves two significant steps: Step 1: decomposes the original cDSP template into two separate cDSP templates, and Step 2: link the two separate cDSP templates using an *Interface*.

The process of the first step is shown in Fig. 4.10. The original cDSP is decomposed into two cDSPs, namely, *cDSP 1* and *cDSP 2*, with the former relating the end mechanical properties as a function of microstructure factors ($D_\alpha, X_f, S_0, Mn, Si, N$) after cooling. The combination of microstructure factors after cooling that best satisfies the end requirements is identified by exercising this sub-cDSP. Regarding *cDSP 2*, it has the best combination of microstructure factor values after the cooling identified from *cDSP 1* as goals. Using *cDSP 2*, the relationship between the microstructure factors after cooling with the microstructure after rolling and the cooling stage operating parameters (AGS, CR, C, Mn) is established. To realize the decomposition, the modification of the original cDSP is as follows.

- For *cDSP 1*:
  - Set ferrite grain size (D_α), phase fraction of ferrite (X_f), pearlite interlamellar spacing (S_0), manganese concentration ([Mn]), the composition of Si ([Si]), and the composition of N ([N]), which are system constraints of the original cDSP, to be system variables.
  - Keep the other constraints and goals the same as the original cDSP.
- For *cDSP 2*:

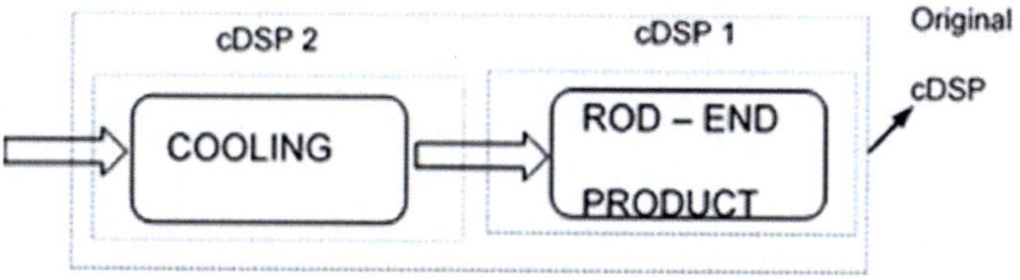

Fig. 4.10 Decomposition of the original design cDSP

- Keep the system variables, namely, Cooling Rate (CR), Austenite Grain Size (AGS), the carbon concentration ($[C]$), and the manganese concentration ($[Mn]$), the same as they are in the original cDSP.
- Set ferrite grain size (D_α), phase fraction of ferrite (X_f), and pearlite interlamellar spacing (S_0), which are system variables of *cDSP 1*, to become system goals.
- Set the final values of D_α, X_f, and S_0 obtained from *cDSP 1* to be the targets of the system goals of *cDSP 2*.

The first cDSP reads as follows:

Given

(1) End requirements identified for the rod rolling process

- Maximize Yield Strength (Goal)
- Maximize Tensile Strength (Goal)
- Maximize Ferrite Fraction (Goal)—Minimize Banding
- Minimize ITT (Requirement)
- Maximize Hardness (Requirement)

(2) Well-established empirical and theoretical correlations, RSMs, and information flow from the end of cooling to the end product mechanical properties (for a further description, see Refs. [14, 17])
(3) System variables and their ranges for cDSP 1 (Table 4.4)

Find
System Variables
X_1, ferrite grain size (D_α)
X_2, the phase fraction of ferrite (X_f)
X_3, the pearlite interlamellar spacing (S_0)
X_4, manganese concentration ($[Mn]$)
X_5, the composition of Si ($[Si]$)
X_6, the composition of N ($[N]$)

Table 4.4 System variables and ranges for cDSP [2]

Sr. No.	System variables	Ranges
1	X_1, ferrite grain size (D_α)	5–25 μm
2	X_2, the phase fraction of ferrite (X_f)	0–1
3	X_3, the pearlite interlamellar spacing (S_0)	0.15–0.25 μm
4	X_4, manganese concentration ($[Mn]$)	0.2–1.5
5	X_5, the composition of Si ($[Si]$)	0.18–0.3
6	X_6, the composition of N ($[N]$)	0.007–0.009

Deviation Variables
$d_i^-, d_i^+, i = 1,2,3$
Satisfy
System Constraints

- Minimum yield strength constraint
- Maximum yield strength constraint
- Minimum tensile strength constraint
- Maximum tensile strength constraint
- Minimum hardness constraint
- Maximum hardness constraint
- Minimum ITT constraint
- Maximum ITT constraint

System Goals
Goal 1:

- Maximize Yield Strength

$$\frac{YS(X_i)}{YS_{Target}} + d_1^- - d_1^+ = 1$$

Goal 2:

- Maximize Tensile Strength

$$\frac{TS(X_i)}{TS_{Target}} + d_2^- - d_2^+ = 1$$

Goal 3:

- Maximize Ferrite Fraction

$$\frac{X_f(X_i)}{X_{f\,Target}} + d_3^- - d_3^+ = 1$$

Variable Bounds
Defined in Table 4.4
Bounds on deviation variables

$$d_i^-, d_i^+ \geq 0 \text{ and } d_i^- * d_i^+ = 0,\ \text{i} = 1,2,3$$

Minimize
We minimize the deviation function

$$Z = \sum_{i=1}^{3} W_i\left(d_i^- + d_i^+\right); \sum_{i=1}^{3} W_i = 1$$

The results obtained after exercising the *cDSP 1* are provided in Refs. [14, 17] and are not repeated here. We understand from the results that a low ferrite grain size and low pearlite interlamellar spacing enhance the product's end mechanical properties. From the results obtained, we also learn the influence of high ferrite, high pearlite, and banded microstructures after cooling on end mechanical properties [14, 17]. The second cDSP is formulated next, with the results from the first cDSP as target goals. The second cDSP is as follows:

Given

(1) Target values for microstructure after cooling (the best combination identified from the first cDSP)
(2) Well-established empirical and theoretical correlations, RSMs, and complete information flow from the end of rolling to the end product mechanical properties (more description see Refs. [14, 17])
(3) System variables and their ranges for cDSP 2 (Table 4.5)

Find
System Variables
X_1, Cooling Rate (CR)
X_2, Austenite Grain Size (AGS)
X_3, the carbon concentration ($[C]$)
X_4, the manganese concentration ($[Mn]$)
Deviation Variables
$d_i^-, d_i^+, i = 1, 2, 3$
Satisfy
System Constraints

Table 4.5 System variables and ranges for cDSP [2]

Sr. No	System variables	Ranges
1	X_1, Cooling rate (CR)	11–100 K/min
2	X_2, Austenite grain size (AGS)	30–100 μ m
3	X_3, the carbon concentration ($[C]$)	0.18–0.3%
4	X_4, the manganese concentration ($[Mn]$)	0.7–1.5%

- Minimum ferrite grain size constraint
- Maximum ferrite grain size constraint
- Minimum pearlite interlamellar spacing constraint
- Maximum interlamellar spacing constraint
- Minimum ferrite phase fraction constraint (manage banding)
- Maximum ferrite phase fraction constraint (manage banding)
- Minimum manganese concentration constraint (manage banding)
- Maximum manganese concentration constraint (manage banding)
- Maximum carbon equivalent constraint (manage banding)

System Goals
Goal 1:

- Minimize Ferrite Grain Size (Achieve Target)

$$\frac{D_{\alpha Target}}{D_{\alpha}(X_i)} + d_1^+ - d_1^- = 1$$

Goal 2:

- Minimize Pearlite Interlamellar Spacing (Achieve Target)

$$\frac{S_{oTarget}}{S_o(X_i)} + d_2^+ - d_2^- = 1$$

Goal 3:

- Maximize Ferrite Fraction (Achieve Target)

$$\frac{X_f(X_i)}{X_{fTarget}} + d_3^- - d_3^+ = 1$$

Variable Bounds
Defined in Table 4.5
Bounds on deviation variables

$$d_i^-, d_i^+ \geq 0 \text{ and } d_i^- * d_i^+ = 0, \text{ i} = 1,2,3$$

Minimize
We minimize the deviation function

$$Z = \sum_{i=1}^{3} W_i\left(d_i^- + d_i^+\right); \sum_{i=1}^{3} W_i = 1$$

The results obtained after exercising the *cDSP 2* are provided in Refs. [14, 17] and thus are not repeated here. The connection between *cDSP 1* and *cDSP 2* is that the output, i.e., the final values of the system variables, of *cDSP 1* comprises the input, i.e., the targets of the system goals, of *cDSP 2*. This connection represents the information workflow linking two cDSPs, which maps to the earlier Step 2 for editing the original cDSP template. The editing and associated consistency checking processes on the platform are shown in Fig. 4.11. The template editor can instantiate two new cDSP templates on the canvas, as highlighted by two red rectangles marked as "End Product cDSP" and "Cooling cDSP" representing *cDSP 1* and *cDSP 2*, respectively. These two cDSP templates' instantiation is the same as the one illustrated in Fig. 4.11. It is noted that many modules of the original cDSP template are reused due to the modularization during the instantiation process of the two new cDSP templates. The link between two cDSP templates is captured by instantiating an Interface marked as "Exchange" highlighted in the circle. Configuration of the Interface is performed in the right window, where information in terms of an interface type, strength, and information flow is specified. According to the interaction between the two cDSP templates, the information flow is weak (one-way), sequential, and flows from *cDSP 1* to *cDSP 2*. The content of the flow is the values of the five system variables of *cDSP 1*. Before executing the edited decision templates, the Editor needs to check any inconsistency due to editing. The consistency checking process is shown in the panel at the bottom of Fig. 4.11. Consistency rules can be dynamically defined and added into the reasoner for reasoning. The newly edited cDSP templates would be ready for execution, storage, and reuse if no rule is violated.

4.9 Variant Design

The Template Implementer makes parametric modifications to the already developed decision templates in a variant design and executes the templates for different scenarios. In this chapter, we showcase a variant design by executing the cDSP template of the original design for different scenarios identified by assigning weights to the deviations associated with each goal. We also illustrate the efficacy of the ternary plots in PDSIDES to support the Template Implementer in exploring the solution space of variant designs to make appropriate design decisions. For the problem formulated in the original cDSP, the Template Implementer is interested in accomplishing the following goals: maximizing ferrite fraction (to manage banding), maximizing tensile strength, maximizing yield strength, and minimizing impact transition temperature. To visualize the goals in the ternary space, the Template Editor must first edit the original cDSP template to remove the goal on impact transition temperature

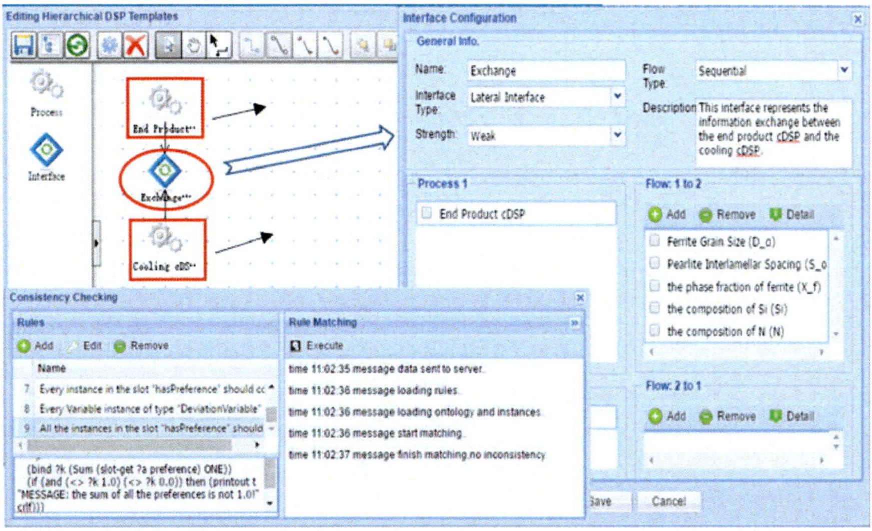

Fig. 4.11 Editing the HRRS design decision template in PDSIDES [2]

and assign it as a constraint with a minimum and maximum value. This is because it is known that the impact transition temperature is directly influenced by changes in weights to other goals and hence need not be considered as an explicit goal. Thus, the variant design cDSP has three goals: maximizing ferrite fraction, tensile strength, and yield strength. Having developed the variant design cDSP, the next step for the template implementer is to identify the design scenarios for execution.

On the platform, the identification of design scenarios is facilitated by the panel shown in Fig. 4.12. The template implementer can specify several weight combinations (each combination stands for one scenario) for goal deviations using the table on the top. Then, PDSIDES will calculate the result concerning each of the weight combinations. In this example, 19 different scenarios are identified (the reader is referred for further information on identifying scenarios to [12]). The template implementer exercises the original cDSP template in variant design scenarios, and the results obtained are sent to MATLAB (at the back-end of PDSIDES) to plot the ternary plots shown in the bottom panel of Fig. 4.12. The template implementer identifies the regions (weight combinations) that satisfy the requirements from the ternary plots. More information on the creation of the ternary plots and their evaluation is available in [12].

The ternary plot regarding the ferrite fraction is presented in Fig. 4.13. The requirement for the template implementer is to maximize ferrite fraction to a value of 0.8, while the maximum value achieved on exercising the cDSP is 0.7116 identified by the light blue dashed line in the red contour region of Fig. 4.13. Any weight combination of goals in this region achieves a high ferrite fraction. Similarly, the high pearlite fraction region is identified by the blue region in Fig. 4.13. The highly banded ferrite-pearlite microstructure region is identified in the boundary between

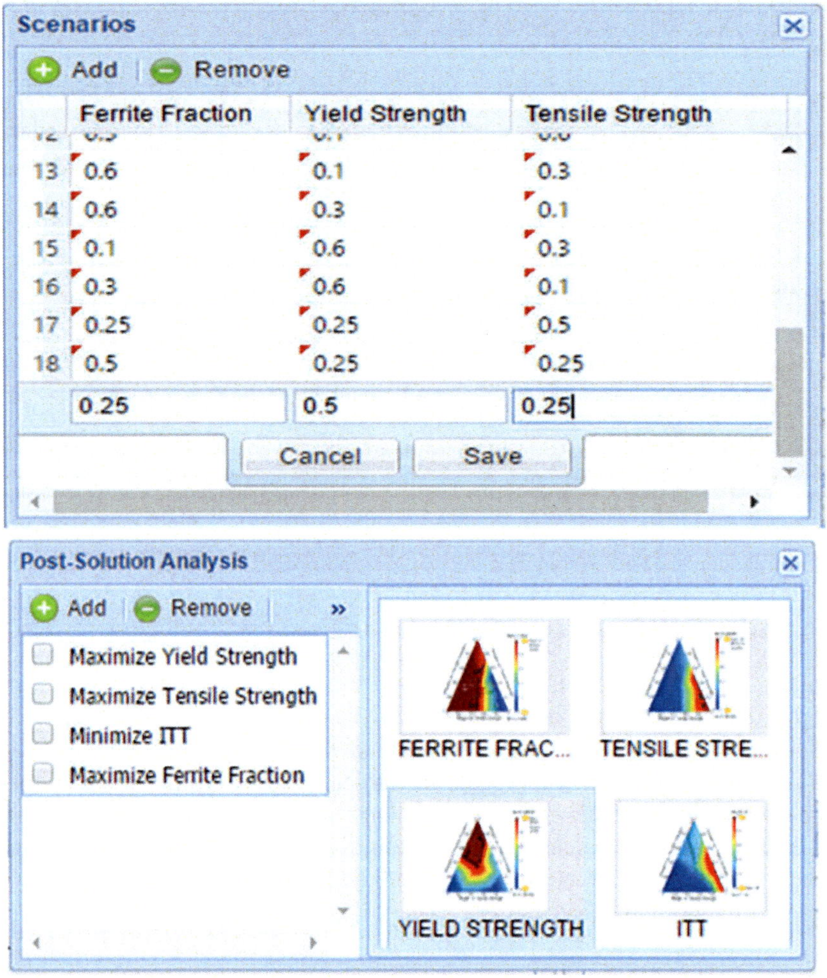

Fig. 4.12 Exercising the HRRS design decision template in PDSIDES [2]

these two regions. The same method is extended to identify the regions that satisfy tensile strength, yield strength, and impact transition temperature requirements.

In Fig. 4.14, we showcase the ternary plot for tensile strength. The requirement for the designer is to maximize tensile strength to a value of 780 MPa. The maximum achieved on exercising the cDSP is 662 MPa and a satisfactory region is identified by the green dashed line in the red contour region of Fig. 4.14. Any weight combination in this region satisfies this requirement.

In Fig. 4.15, we present the ternary plot for yield strength. The requirement for the designer is to achieve a maximum tensile strength of approximately 330 MPa. The maximum value achieved on exercising the cDSP is 284 MPa and a satisfactory value is identified by the dark blue dashed lines in the red contour region of Fig. 4.15.

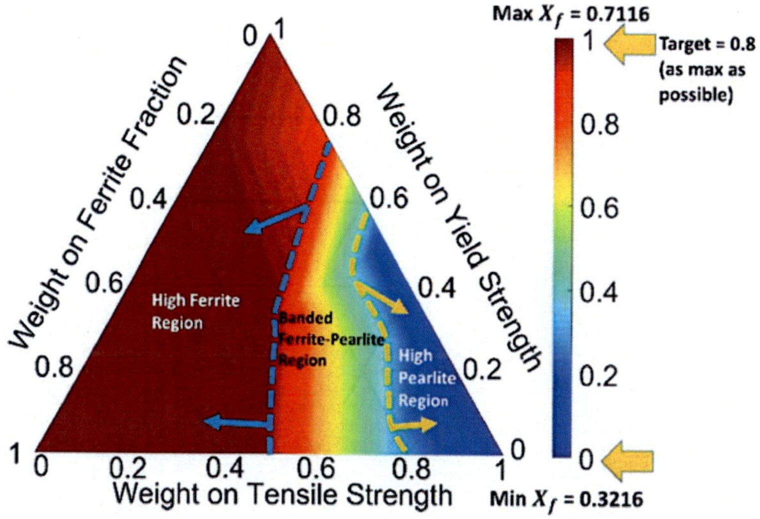

Fig. 4.13 Ternary plot for ferrite fraction

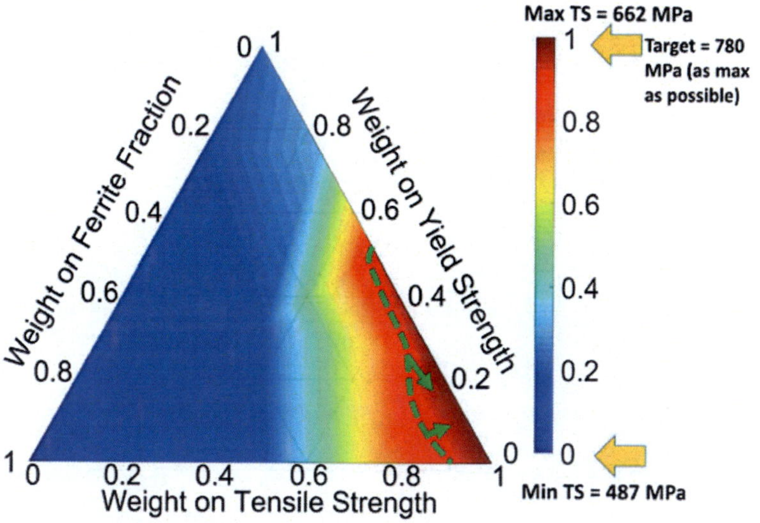

Fig. 4.14 Ternary plot for tensile strength

In Fig. 4.16, we illustrate the ternary space for achieving the impact transition temperature (ITT) when weights are changed for the three goals.

The requirement for the designer is to achieve a minimum impact transition temperature. The minimum value achieved is −42 °C, a sastifactory region is identified by the red dashed line in the dark blue contour region of Fig. 4.16. The maximum ITT achieved is 99 °C and is identified by the red contour region.

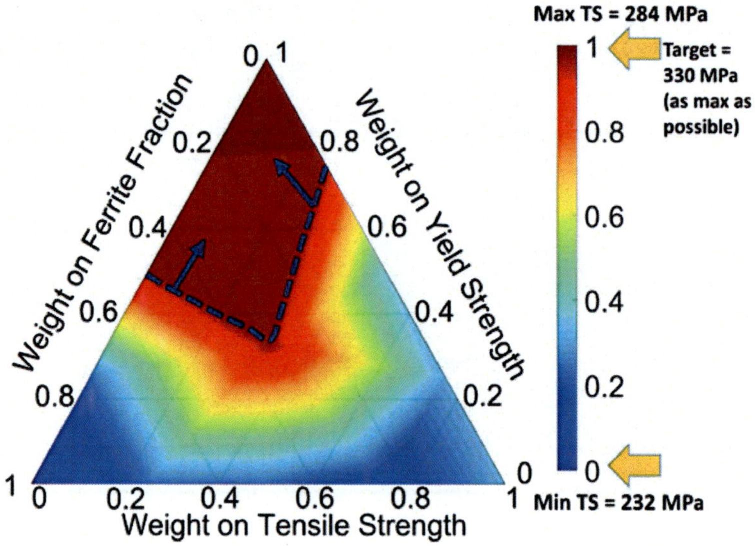

Fig. 4.15 Ternary plot for yield strength

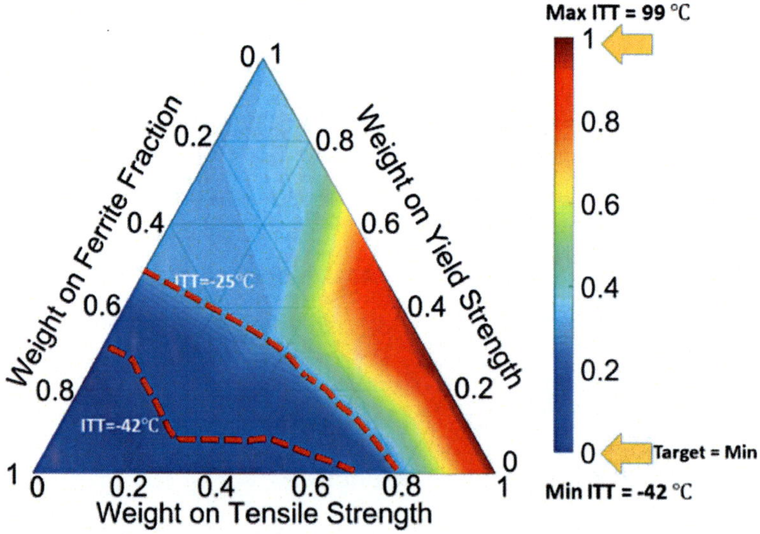

Fig. 4.16 Ternary plot for ITT

Since the template implementer's interest is to identify a common region that satisfies all goals, a superposed ternary plot having all the goals is shown in Fig. 4.17. Several solution weight points from the superposed ternary plot (A, B, C, D, E, F, G) are identified and analyzed. The results associated with these solution weight points are summarized in Table 4.6.

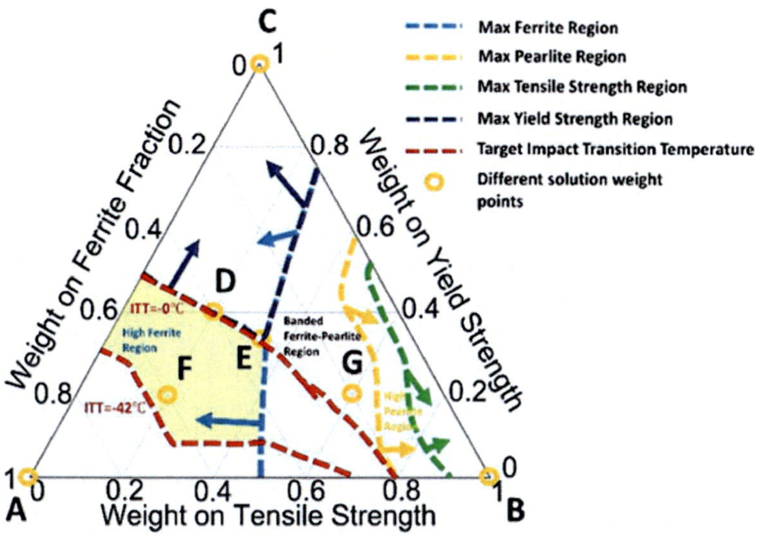

Fig. 4.17 Superposed ternary plot

Table 4.6 Identified solution points after exploration

Sol. Pt	CR K/min	AGS (μm)	C (%)	Mn (%)	X_f	YS MPa	TS MPa	ITT °C
A	16.5	99.9	0.18	0.7	0.71	232	487.7	-26
B	99.9	30	0.29	1.5	0.32	248	662	99
C	22.8	30	0.18	1.5	0.7	284	526	3.5
D	11	30	0.18	1.5	0.71	283	519	0
E	11	30	0.18	1.5	0.71	283	519	0
F	11	30	0.18	0.7	0.7	244	513	-42
G	62	30	0.19	1.5	0.65	281	547	15

Analyzing Fig. 4.17 and Table 4.6 indicates that the light-yellow region satisfies all requirements for managing banding (high ferrite), maximizing yield strength, maximizing tensile strength, and minimizing ITT in the best possible manner. However, the requirements for high tensile strength and high yield strength are compromised to satisfy the requirements of managing banding and minimizing ITT. It is also observed that a high ferrite region supports maximizing yield strength and minimizing ITT. However, maximizing the tensile strength is supported by a high pearlite fraction. Point F out of these multiple solutions listed in Table 4.6 is chosen as the solution as F satisfies all the requirements in the best possible manner.

By reusing the knowledge archived in the original HRRS design cDSP template for execution and utilizing the ternary plot for post-solution analysis, the template implementer explores the solution space of variant designs and makes appropriate design decisions.

4.10 Validation of PDSIDES

4.10.1 Empirical Structural Validation

Empirical structural validation involves accepting the appropriateness of the example problem used to verify the performance of the PDSIDES for original, adaptive, and variant designs. In this chapter, the gear manufacturing-process design problem focused on rod rolling—a complex system design that calls for a series of decisions to be made—is introduced. Decisions to be made at each manufacturing unit are formulated as cDSPs and linked as a decision network. In original design addressed in Sect. 4.7, the template creator (domain expert) formulates in PDSIDES, the cDSP for the problem boundary framed within the hot rod rolling process chain problem by taking into account the complete information flow across models thereby establishing relationships. Using the formulated cDSP the ability of the PDSIDES platform to be used to carry out original design is demonstrated. In adaptive design addressed in Sect. 4.8, the template editor (senior designer) modifies the existing original design cDSP template to satisfy new requirements. These requirements can be easily satisfied by editing the existing formulated original design cDSP template in PDSIDES. The editing involves two major steps: Step 1, decompose the original cDSP template into two separate cDSP templates, and Step 2, link the two separate cDSP templates using an *Interface*. Two cDSPs are formulated from the original design cDSP to demonstrate adaptive design. The cDSPs are interlinked via an interface of design variables that are shared. Using the cDSPs formulated, the ability of the PDSIDES platform to carry out adaptive design is demonstrated.

4.10.2 Empirical Performance Validity

Empirical performance validation consists of accepting the usefulness of the outcome with respect to the initial purpose and accepting that the achieved usefulness is related to using PDSIDES for original, adaptive and variant designs. In PDSIDES, decision-related knowledge is modeled as modular, computational templates based on the DSP constructs using ontology to facilitate execution and reuse. The advantages of PDSIDES are that it provides the functionality to capture knowledge when Template Creators create decision templates in original design, maintain consistency when Template Editor modify decision templates in adaptive design and provide a package of documented knowledge when Template Implementers executes decision templates in variant design.

4.11 Role of Chapter 4 and Remarks on the Knowledge-Based Platform PDSIDES

Engineering system design is fundamentally a decision-making process and knowledge plays a critical role in facilitating decision-making. In this chapter, a Knowledge-Based Platform for Decision Support in the Design of Engineering Systems is presented and validated using a steel manufacturing-process chain problem. In PDSIDES, decision-related knowledge is modeled with modular, computational templates based on the DSP constructs using ontology to facilitate execution and reuse. In order to provide users of different knowledge levels with a proper decision support, we define three types of users, namely, Template Creators, Template Editors, and Template Implementers, who perform original design, adaptive design, and variant design respectively. The unique advantage of PDSIDES is that it provides the functionality to capture knowledge when Template Creators create decision templates in original design, maintain consistency when a Template Editor modifies decision templates in adaptive design and provides a package of documented knowledge when Template Implementers execute decision templates in variant design.

Distributed information control is not yet considered in the current version of PDSIDES. Future research opportunities lie in enabling the negotiation of collaborative decisions that are controlled by different stakeholders. For example, in the HRRS design example process designers at different stages such as rolling, and cooling may not be willing to sharing the full information in their own decision-making process, and then the negotiation of a collaborative decision is needed. Providing the functionality for negotiating collaborative decisions would be valuable for the application of PDSIDES in a supply chain environment, where the decision-makers are distributed. All these can be addressed by enabling the PDSIDES platform in the cloud (Cloud-based PDSIDES) [23].

4.12 Where We Are and What Comes Next?

In this chapter, we present the design and implementation of the PDSIDES platform. In the design part, we define three user types and their specific working scenarios. We also define the knowledge-based decision support modes for different users and working scenarios. In the implementation part, we introduce PDSIDES' architecture, components, programing language, and GUI portal. In the following chapters (Chaps. 5 and 6), we add more functionalities to PDSIDES to provide knowledge-based support in the meta-design of decision workflows and robust design space exploration under uncertainty.

References

1. Mocko, G. M., Rosen, D. W., & Mistree, F. (2007). Decision retrieval and storage enabled through description logic. In: *ASME Computers and Information in Engineering Conference, September 4–7, Las Vegas, Nevada. Paper Number: DETC2007-35644.*
2. Ming, Z., Nellippallil, A. B., Yan, Y., Wang, G., Goh, C. H., Allen, J. K., & Mistree, F. (2018). PDSIDES—A knowledge-based platform for decision support in the design of engineering systems. *Journal of Computing and Information Science in Engineering, 18*(4), 041001.
3. Baeza-Yates, R., & Ribeiro-Neto, B. (2011). *Modern information retrieval: The concepts and technology behind search.* Addison Wesley.
4. Salton, G., Wong, A., & Yang, C.-S. (1975). A vector space model for automatic indexing. *Communications of the ACM, 18*(11), 613–620.
5. Jess, the Rule Engine for the JavaTM Platform, 2015. http://herzberg.ca.sandia.gov/
6. Sencha GXT. https://www.sencha.com/products/gxt/#overview
7. Google Web Toolkit. http://www.gwtproject.org/overview.html
8. Apache Flex. http://www.adobe.com/devnet/flex.html
9. JSON. JavaScript Object Notation. http://www.json.org/
10. Allen, J. K., Mistree, F., Panchal, J., Gautham, B., Singh, A., Reddy, S., Kulkarni, N., & Kumar, P. (2015). Integrated realization of engineered materials and products: A foundational problem. In *Proceedings of the 2nd World Congress on Integrated Computational Materials Engineering* (pp. 277–284). Wiley Online Library.
11. Nellippallil, A. B., Rangaraj, V., Gautham, B. P., Singh, A. K., Allen, J. K., & Mistree, F. (2017). A goal-oriented, inverse decision-based design method to achieve the vertical and horizontal integration of models in a hot-rod rolling process chain. In *ASME Design Automation Conference,* Paper Number DETC2017-67570.
12. Nellippallil, A. B., Song, K. N., Goh, C.-H., Zagade, P., Gautham, B., Allen, J. K., & Mistree, F. (2017). A goal-oriented, sequential, inverse design method for the horizontal integration of a multi-stage hot rod rolling system. *Journal of Mechanical Design 139*(3), 031403.
13. Nellippallil, A. B., Song, K. N., Goh, C.-H., Zagade, P., Gautham, B., Allen, J. K., & Mistree, F. (2016). A goal oriented, sequential process design of a multi-stage hot rod rolling system. In *Proceedings of the ASME 2016 International Design Engineering Technical Conferences and Computers and Information in Engineering Conference,* Paper Number DETC2016-59402.
14. Nellippallil, A. B., Rangaraj, V., Gautham, B., Singh, A. K., Allen, J. K., & Mistree, F. (2018). An inverse, decision-based design method for integrated design exploration of materials, products, and manufacturing processes. *Journal of Mechanical Design, 140*(11), 111403.
15. Nellippallil, A. B., Mohan, P., Allen, J. K., & Mistree, F. (2019). Inverse thermo-mechanical processing (ITMP) design of a steel rod during hot rolling process. *Proceedings of the ASME 2019 International Design Engineering Technical Conferences and Computers and Information in Engineering Conference,* Paper Number DETC2018-85913.
16. Nellippallil, A. B., Mohan, P., Allen, J. K., & Mistree, F. (2020). An inverse, decision-based design method for robust concept exploration. *Journal of Mechanical Design, 142*(8), 081703.
17. Nellippallil, A. B., Allen, J. K., Gautham, B., Singh, A. K., & Mistree, F. (2020). Integrated design exploration of materials, products, and manufacturing processes using goal-oriented, inverse design method. In *Architecting robust co-design of materials, products, and manufacturing processes* (pp. 207–262). Springer.
18. Nellippallil, A. B., Vignesh, R., Allen, J. K., Mistree, F., Gautham, B. P., & Singh, A. K. (2017). A decision-based design method to explore the solution space for microstructure after cooling stage to realize the end mechanical properties of hot rolled product. In: *4th World Congress on Integrated Computational Materials Engineering (ICME 2017).*
19. Kuziak, R., Cheng, Y.-W., Glowacki, M., & Pietrzyk, M. (1997). Modeling of the microstructure and mechanical properties of steels during thermomechanical processing. *NIST Technical Note(USA), 1393,* 72.
20. Yada, H. (1987). Prediction of microstructural changes and mechanical properties in hot strip rolling. *Accelerated Cooling of Rolled Steel,* 105–119.

21. Gladman, T., McIvor, I., & Pickering, F. (1972). Some aspects of the structure-property relationships in high-C ferrite-pearlite steels. *Journal of the Iron and Steel Institute, 210*(12), 916–930.
22. Hodgson, P., & Gibbs, R. (1992). A mathematical model to predict the mechanical properties of hot rolled C-Mn and microalloyed steels. *ISIJ International, 32*(12), 1329–1338.
23. Nellippallil, A. B., Ming, Z., Allen, J. K., & Mistree, F. (2019). Cloud-based materials and product realization—Fostering ICME via Industry 4.0. *Integrating Materials and Manufacturing Innovation, 8*(2), 107–121.

Chapter 5
Knowledge-Based Meta-Design of Decision Workflows

Utilizing an enterprise's capital related to the knowledge of the design processes has become crucial to improving the enterprise's agility and ability to respond to market shifts or changes. The complexity and uncertainty of design processes raise the challenge of capturing tacit knowledge and aid in designing design processes. This chapter involves capturing, representing, and documenting knowledge related to hierarchical decision workflows in the meta-design of complex engineered systems. The ontology is developed in the context of the Decision Support Problem Technique (DSPT), considering the requirements of guiding assistance in designing design workflows and integrating the problem, product, and process information in a design decision-making process. The meta-design of a heat exchanger in a small thermal system is presented as an example to illustrate the effectiveness of the knowledge model. A summary of this chapter is presented in Table 5.1. The mapping of the sections to the components (topics) discussed in this chapter is presented in Row 2 of Table 5.1.

5.1 Frame of Reference

Ontology is defined as a specification of a conceptualization, which can provide a shared vocabulary for representing domain-specific knowledge [1]. Ontology has a tremendous potential impact on engineering system designs [2], with the expected benefits of employing ontologies being [3–5]:

- Flexibility—knowledge is defined in terms of an ontology instead of "hardcoding" within the platform
- Intelligent behavior—knowledge can be derived from the factual knowledge explicitly represented in the ontologies
- Semantic interoperability—semantics of the (possibly several) languages used by the platform's external parties can be defined by a set of inter-related ontologies

© The Author(s), under exclusive license to Springer Nature Switzerland AG 2022
Z. Ming et al., *Architecting A Knowledge-Based Platform for Design Engineering 4.0*,
https://doi.org/10.1007/978-3-030-90521-7_5

Table 5.1 Summary interpretation of the problem investigated in this chapter

Elements	What?
Components	**Requirement of modeling meta-design of decision workflows (Sect. 5.2)**, ontology development (Sect. 5.3), testing the performance of the ontology (Sect. 5.4)
Connectors	Meta-design and ontology
Form	**How?**
Component roles	Identify and capture the knowledge used for meta-design of design processes
Properties	Requirements of developing the meta-design ontology
Relationship	The PEI-X diagram
Rationale	**Why?**
Motivation	To build a knowledge based for decision-based meta-design and provide knowledge support to assist designers in process design
Assumptions	Any design process can be represented using phases, events, and information
Interpretation	Understanding how meta-design related knowledge is represented using ontology

- Expressiveness—context information is represented using a formal representation language, which automatically checks the consistency of the models.

Based on the different classifications of the knowledge in design, ontologies are categorized into "domain" and "task" ontology [6]. The former is concerned with the "design target", i.e., the product-related knowledge such as the product ontology for PLM, service ontology for product and service family design, and device ontology. Process-related knowledge is represented by "task" ontologies, which focus on the semantic process modeling for a problem-solving method and manufacturing operation, e.g., Process Specification Language (PSL), from the National Institute of Standards and Technology (NIST) [7], and TOVE (Toronto Virtual Enterprise) for business and enterprise [8]. Several ontologies for design process knowledge have been developed, taking into account various design views and functions. However, the knowledge representation related to the decision-making process needs to be further studied and compared with the business processes and generic engineering design activities. For example, in [8–12], different ontologies are proposed to improve the analysis, operation, and management of business processes, while in [7, 13–15], a generic set of design activities is identified, and the corresponding ontologies are developed to achieve a shared understanding and consistency of these activities. In contrast, the defined decision support ontologies mainly focus on the description of the decision-making itself while ignoring the significant tacit knowledge of organizing the decision-making activities from the perspective of system design that involves the partitioning and planning of the appropriate decision support problems. Consequently, the previous knowledge of decision-making activities is challenging to reuse in designing new complex engineering systems.

Some studies begin to explore the integration of design information by defining the top-level ontology (upper ontology or foundation ontology) or select an available upper ontology, e.g., Basic Formal Ontology (BFO), General Formal Ontology (GFO), or Standard Upper Merged Ontology (SUMO). For instance, Kitamura and Mizoguchi [6] define a systematized functional ontology for the knowledge modeling of artifacts in the conceptual design phase, Štorga et al. [16] design an ontology based on SUMO, which achieves a general design description for product development. In this paper, we focus on defining a domain-independent ontology to capture, represent, and document knowledge in meta-design process hierarchies from the perspective of DBD.

5.2 Requirements for Meta-Design Process Hierarchies Model

From the DBD perspective, meta-design is a meta-level process in designing systems, including partitioning the design problem, decomposing the design processes into a set of decisions, and planning the decision-making sequence [17]. In the context of DSPT, a meta-design outputs an SPs word formulation and an initial plan, i.e., an initial solution network in the form of DSPT Palette entities. In the proposed platform PDSIDES, the meta-design process is modeled computationally to facilitate designers formulating a hierarchy of decision workflows and solving the DSPs. In keeping with the feature of design systems/processes hierarchy, as well as considering the needs of designers as users of PDSIDES in using the DSPT to designing decision workflows, we identify the requirements for the computational model of meta-design process hierarchies as follows:

- **Decision-Centric Process Design**

DBD is a mindset emphasizing the principal function of designers in design is to make decisions. In the context of DBD, design involves a series of decision activities, with some of them made sequentially while others concurrently. Those design decisions represent a form of communication with the features in terms of being domain-dependent and domain-independent. By focusing on the decisions, the design processes can be described in a common "language" used by teams from various disciplines while designing the design processes. Design decisions involve some characteristics associated with the design of engineering systems that are summarized in [17] and are equally applicable to different design processes in the life cycle.

- **Flexible Decomposition and Reconfiguration**

Engineering systems, e.g., ships and airplanes, are often too complex to be handled in their entirety, and it is necessary to decompose these systems into smaller interrelated subsystems. This hierarchical system feature determines the multileveled and multidimensional designing of engineering systems [18]. Multilevel design refers to designing decision workflows on a physical level, i.e., define the "parent" system/process for other subsystems/subprocesses. In contrast, multidimensional design involves designing decision workflows and a timeline, i.e., defining a stage in the process as an action-specific activity. It requires a decomposable and solvable model to capture the complexity of the engineered systems using the available computational power.

Furthermore, the model also needs to support dynamic decomposition as the design evolves. Meanwhile, reducing the number of iterations and reusing the existing processes, e.g., adaptive design and variant design, are also critical requirements for designing process. Therefore, a flexible and reconfigurable design process is necessary.

- **Information Separation and Synthesis**

The DSPT is a concrete implementation of the decision-based design philosophy. In DSPT, the design is regarded as an information transformation process, which converts the information of the product needs into knowledge about the product [17]. Therefore, the purpose of decision workflows is information transformation. Currently, traditional tools, e.g., FIPER, ModelCenter, and iSIGHT, capture the information of the design processes to integrate the specific product information, which lacks supporting meta-design in PLM frameworks [19]. Thus, it is hard to reuse the definitions of different design processes to design a new product. In [20], a method for resolving this issue of reusability by using domain-independent process templates is proposed. Process templates support information transformations and embody the interactions among these transformations. The template-based method depends on developing reusable templates via separating procedural and declarative information [21]. Procedural information is the information about the transformation process, while declarative information refers to the information associated with product status. Additionally, it is equally necessary to consider the activity synthesis, as designers need to generate various ideas, which are analyzed and synthesized into forms that may represent acceptable solutions to the problem.

- **Information Interaction**

Meta-design involves information interactions at different levels and dimensions. From the perspective of the PDSIDES users, it is necessary to express information about design decisions in a standard format to facilitate designers' interactions [22]. From the perspective of the PDSIDES platform, the design information model based on the DSPT involves the interactions of the model associated with the design processes. Model interactions include the interactions among the various subsystems/subprocesses and between the "parent" system/process and subsystems/subprocesses. Furthermore, due to the system's characteristics discussed above, the interaction among product, process, and problem must also be considered.

- **Knowledge Reuse**

With the progress of design processes, the freedom to make decisions is reduced, and the qualitative ratio of hard and soft information increases [23]. This means that the degree of partitioning and planning in a meta-design needed to proceed as the design increases. For example, in DSPT, the reuse of cross-domain design information is established by increasing the ratio of hard to soft information, especially in the early design stages. Designers can reuse existing knowledge to influence idea generation for design problem solutions and understanding how these ideas were evaluated or even reusing the problem-solving methods.

- **Visualization**

Designing decision workflows require the development process of the model to be visualized. The model of meta-design process hierarchies is identified as a complex system with the characteristic of interactions among the "parent" system/process and the subsystems/subprocesses. Designers will have intuitive awareness and understanding of their design problem through hierarchical structural visualization of the computational model. This will increase the efficiency of model editing.

5.3 Ontology Development for Designing Decision Workflows

Based on the requirements identified, in this section we present an ontology for the computational model of the meta-design process hierarchy. In a computer environment, the ontology is accomplished by supporting the designer's partition activities, i.e., dividing the functions, structures, and processes, splitting the system into several system layers, i.e., subsystems, sub-subsystems, etc. In the meta-design process hierarchy, decomposition and reconfiguration are accomplished via the template-based approach, with its effectiveness demonstrated in [24]. Information separation and synthesis are facilitated via the interaction of information flow among different entities. The information expression related to designing of decision workflows in a standardized format is also required. Therefore, the execution of the process templates is facilitated by employing a standard and computer-interpretable modeling language underlying the ontology, namely OWL [1]. The latter supports parsing through other applications, such as Java function calls, while visualization is achieved via a graph-based editing tool in Protégé [25].

(1) Identification of Concepts and Relations for PEI-X Ontology

Domain concepts are often described by a vocabulary of terms. DSPT Palette entities have identified a set of terms for modeling the process template, including `Phase`, `Event`, `Decision`, `Task`, `System`, `Information`, and `Knowledge` [22]. These terms are adopted and reused here by adapting concepts and explicitly defining them as `Classes` based on the information model. In an ontology, the semantic relationships among concepts are captured using `Slots`. The definitions for the top-level concepts of the PEI-X ontology and related properties are:

- Class ***Process_Template***

Design processes are carried out through various design activities, from the initial requirements to the final product. The `Class` ***Process_Template*** is a general representation of design activity workflows, which convert product information from one stage to another. Within each ***Process_Template***, a network of transformations operates and converts product design information via solving the problems associated with design decision-making.

The concept of the `Class` ***Process_Template*** is shown in Fig. 5.1. This class integrates all the template modules and represents the information structure of a meta-design process in different stages. It represents a standard, scalable hierarchical building block where the solid box stands for the information integration unit with the associated instances of classes ***SPs_Entity***, ***Sys_Entity***, ***GeneralDesign_Knowledge***, and ***DSP_Template*** plugged in (the dashed box is shown in Fig. 5.1). A hierarchy is built by assembling a series of different ***SPs_Entity*** via the `Class` ***Interface*** that has been defined in [26]. The data and object `Slots` for corresponding `Class` ***Process_Template*** in PEI-X ontology are described in Table 5.2.

- Class ***SPs_Entity***

The `Class` ***SPs_Entity*** is a general representation of SPs that consists of a series of sub-classes: ***Phase***, ***Event***, ***Decision***, and ***Task*** based on the definitions of the DSPT Palette entity classes [22]. A ***Phase*** represents pieces of a partitioned process,

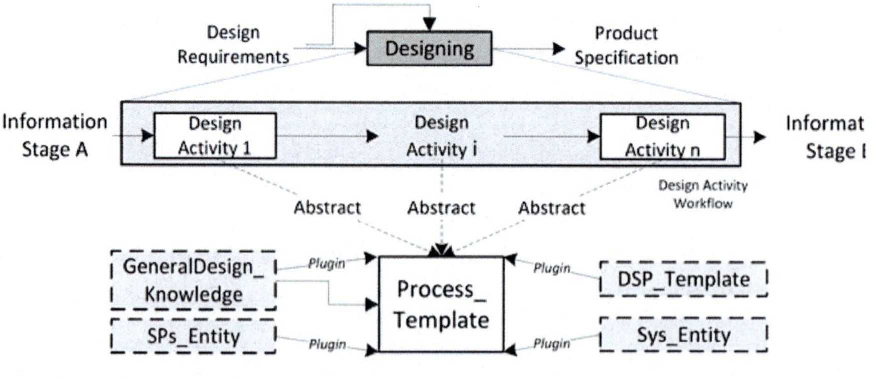

Fig. 5.1 Concept of Class Process_Template

Table 5.2 Slots of Process_Template

Slot name	Definition
templateType	Symbol. Specifies the type of a *Process_Template* (e.g., "Process", "Phase", "Event", "Decision", "Task")
designFor	Specifies the instance of design object that has a certain *System* in *SPs_Entity*
has_SPs	Specifies the instance of associated *SPs_Entity*
use_Knowledge	Specifies the instance of associated *GeneralDesign_ Knowledge*

e.g., "Designing for Concept" and "Designing for Manufacture", an ***Event*** occurs within a phase, and ***Task*** and ***Decision*** are employed to model phases and events. Phases and events are accomplished by performing tasks and making decisions. By instantiating those sub-classes, designers use them to model the timeline for designing a specific system. Different SPs entities correspond to the different levels of the design problem. The support for human designers using the DSP Technique is provided through solutions to Support Problems, especially Decision Support Problems. The concept of the `Class` ***SPs_Entity*** is shown in Fig. 5.2, representing a standard building block where the solid box (or shell) is the information processing unit with the associated instances of `Class` ***Information*** as input and output and lines represent the associated vertical and lateral dependencies.

The data and object `Slots` for corresponding `Class` ***SPs_Entity*** in PEI-X ontology are illustrated in Table 5.3.

- Class ***Sys_Entity***

The `Class` ***Sys_Entity*** is a general representation of a core design object and its related properties. To ensure the concepts' consistency and standardization, here, we employ the revised version of the Core Product Model (CPM) presented by NIST [27], which is conceived as a representation of product development information. The diagram of the `Class` ***Sys_Entity*** is shown in Fig. 5.3, where the `Class` ***Sys_Entity*** involves five sub-classes: ***System***, ***Property***, ***Requirement***, ***Behavior***, and ***Specification***. The key object class in the ***Sys_Entity*** is a ***System*** representing a distinct entity in a product, and its sub-entity is a `Class` ***Component*** and `Class` ***Part***. All the latter

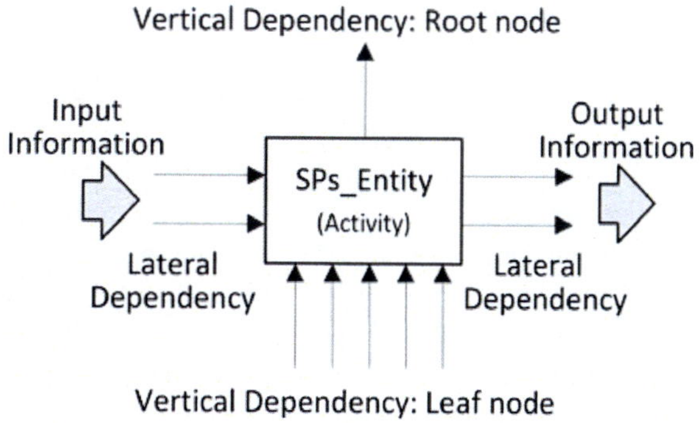

Fig. 5.2 Concept of Class SPs_Entity [28]

Table 5.3 Slots of SPs_Entity

Slot name	Definition
Input	Specifies the input *Information* instance of a *SPs_Entity*
Output	Specifies the output *Information* instance of a *SPs_Entity*
L-Dependency	Specifies the instance of lateral *Interface*
V-Dependency_Root	Specifies the instance of vertical parent *Interface*
V-Dependency_Leaf	Specifies the instance of vertical subsystem *Interface*

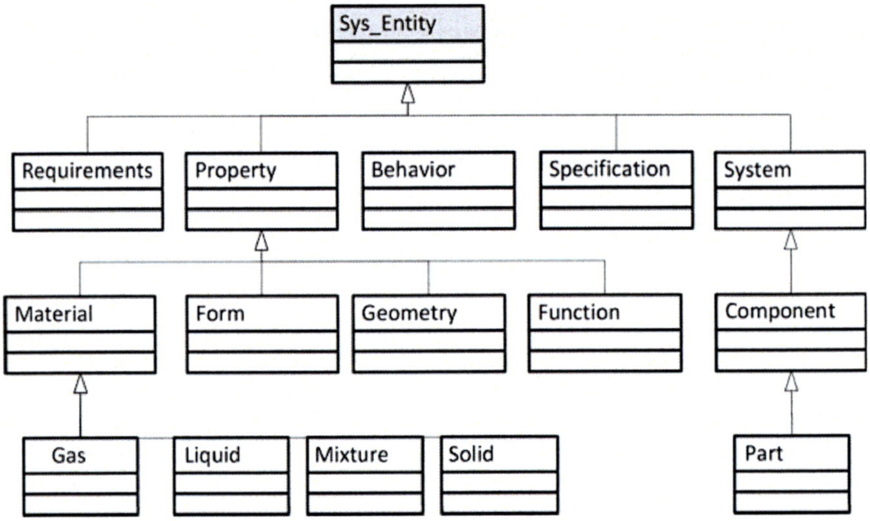

Fig. 5.3 Diagram of Class Sys_Entity [28]

entities can be inter-related through the ***System*** Slots. A ***Property*** is an abstract class from which the `Classes` ***Function***, ***Form***, ***Geometry***, and ***Material*** are specialized. ***The function*** represents one aspect of the system's entity output, while the ***Form*** of the system entity can be viewed as the proposed design solution for the design problem specified by the function. ***Geometry*** is the spatial description of the system entity, and ***Material*** is the description of the internal composition of the system entity. According to the features of the material properties, the `Class` ***Material*** comprises the sub-classes: ***Gas***, ***Liquid***, ***Solid***, and ***Mixture***. A ***Requirement*** is a specific element of the specification of a system entity that governs some aspect of its function, form, geometry, or material according to the customer's needs. A ***Behavior*** represents how the artifact implements its corresponding function. A ***Specification*** describes the collection of information related to the design of a system entity deriving from the customers' needs and the engineering requirements.

The data and object `Slots` for corresponding `Class` ***Sys_Entity*** in PEI-X ontology are illustrated in Table 5.4.

- Class ***GeneralDesign_Knowledge***

The `Class` ***GeneralDesign_Knowledge*** represents hard information derived from a previous design process and can be reused in a new design process. There are several types of design knowledge, such as thc formula for calculations and designer experience, e.g., preference for weight, which will support the solutions for different design activities. The amount of knowledge will be rich as the number of process template instances increases. The data and object `Slots` for the corresponding `Class` ***GeneralDesign_Knowledge*** in a PEI-X ontology are illustrated in Table 5.5.

Table 5.4 Slots of Sys_Entity [28]

Slot name	Definition
Image	String. Specifies the image of an instance
has_Behavior	Specifies the *Behavior* instance of a *Sys_Entity*
has_Form	Specifies the *Form* instance of a *Sys_Entity*
has_Geometry	Specifies the *Geometry* instance of a *Sys_Entity*
has_Material	Specifies the *Material* instance of a *Sys_Entity*
has_EngineeringInformation	Specifies the engineering *Information* instance
has_Requirement	Specifies the *Requirement* instance of a *Sys_Entity*
has_Specification	Specifies the *Specification* instance of a *Sys_Entity*
application	String. Specifies the application of a certain *Behavior* instance
constructional_Features	String. Specifies the constructional features of a certain *Behavior* instance
performance_Features	String. Specifies the performance features of a certain *Behavior* instance
formType	Specifies the instance of *Form*
use_Knowledge	Specifies the instance of associated *GeneralDesign_Knowledge*
density-ρ: kg/m^3	String. Specifies the density value of *Materia*l that is applied to *System*
enthalpy-H: J/mol	String. Specifies the enthalpy value of *Materia*l that is applied to *System*
heatCapacity-C: J/(kg K)	String. Specifies the heat capacity value of *Materia*l that is applied to *System*
tensileStrength-S: kN/m^2	String. Specifies the tensile strength value of *Materia*l that is applied to *System*
thermalConductivity-k: W/(m·K)	String. Specifies the thermal conductivity value of *Materia*l that is applied to *System*
cost-USD/lb	String. Specifies the cost price value of *Materia*l that is applied to *System*

Table 5.5 Slots of GeneralDesign_Knowledge [28]

Slot name	Definition
expression	String. Specifies the expression of a *GeneralDesign_ Knowledge*
input	Specifies the input instance of an *Information*
output	Specifies the output instance of an *Information*

Table 5.6 Slots of information [28]

Slot name	Definition
informationType	Symbol. Specifies the type of *Information* (e.g., "*Concept*", "*Quantity*", "*Function*")
subInformation	Specifies the sub-information of a certain *Information*
attributeOf	Specifies the attributes of a certain *Information* based on the instance of associated Class (e.g., "*Form*", "*Material*", "*System*", "*Requirement*", "*Specification*" "*GeneralDesign_Knowledge*", "*CO: Function*", "*CO: Quantity*", "*CO: Response*", "*CO: Behavior*")
applyTo	Specifies the application of *Information* to a certain *SPs_Entity*

- Class ***Information***

The `Class` ***Information*** represents one of the transmission entities, i.e., information, energy, and matter, to achieve the connections among the entities presented above (***SPs_Entity***, ***Sys_Entity***, and ***GeneralDesign_Knowledge***). All those entities have a common feature requiring some inputs and providing some outputs. The purpose of the information transmission entity is to capture these inputs and outputs. The data and object `Slots` for corresponding `Class` ***Information*** in a PEI-X ontology are illustrated in Table 5.6.

- Reuse the existing classes

In the earlier works of the ontology-based representation of a DSP template, `Classes` ***Interface*** [26], ***CO: cDSP_Template*** and its related sub-classes (see cDSP ontology in [29]), as well as `class` ***SO: sDSP_Template*** and its related sub-classes (see sDSP ontology in [30]) have been reused in the PEI-X ontology.

(2) **Build Meta-Design Process Hierarchies Using PEI-X Ontology**

The definition of `Classes` and associated `Slots` are implemented by using Protégé developed by Stanford University. This tool provides an environment for creating and editing ontologies as well as populating `Instances` based on ontologies. The building procedure and design scenarios for process templates are listed below.

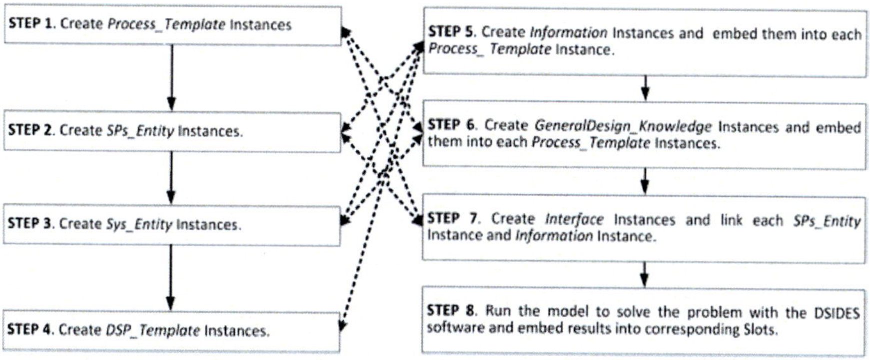

Fig. 5.4 Building Procedure for the Process Templates Hierarchy [28]

- Building Procedure for Process Templates

The definition of `Classes` and associated `Slots` are implemented by using Protégé tool [31] that provides an environment for creating and editing ontologies as well as populating `Instances` based on ontologies. After the development of the ontology, it is essential to create and populate instances. In an ontological environment, most of the populated instances can be reused in future designs. In the PEI-X ontology, as a graphical tool for visual editing of the `Instance` and relationships among `Instances`, the Protégé Graph Widget [25] is used to build the hierarchical design activities in the process templates. To successfully reuse these instances, here we define five types of design scenarios to create process template instances based on their own design goals, namely "Process", "Phase", "Event", "Decision", and "Task". The building procedure is shown in Fig. 5.4.

Step 1. Create ***Process_Template*** `Instances`. Based on the design scenarios, select the relevant blue box marked "Phase", "Event", "Decision", and "Task" from the left panel (represents `Classes`), and drag them to the right canvas (represents `Instances`), and then edit the generated `Instances`. The number of ***SPs_Entity*** `Instance` is determined by specific design problem information.

Step 2. Create ***SPs_Entity*** `Instances`. Based on the ***Process_Template*** `Instance` created in Step 1, double-click the box of ***SPs_Entity*** `Instance` and edit the generated `Instances`.

Step 3. Create ***Sys_Entity*** `Instances`. Based on the ***Process_Template*** `Instance` created from Step 1, create the ***System*** `Instance` and its associated ***Sys_Entity*** `Instances`, then edit the generated `Instances`.

Step 4. Create ***DSP_Template*** `Instances`. Based on the ***Process_Template*** `Instances` created in Step 1, create the appropriate DSP templates (e.g., ***CO: cDSPTemplate*** `Instance` and/or ***SO: sDSPTemplate*** `Instance`) for the solution and edit the generated `Instances`.

Step 5. Create ***Information*** `Instances` and embed them into each ***Process_Template*** `Instance`. Based on the `Instances` created in Steps

1–4, create the corresponding ***Information*** `Instances` for each associated `Instances` based on the relations, and edit the generated `Instances`.

Step 6. Create ***GeneralDesign_Knowledge*** `Instances` and embed them into each ***Process_Template*** `Instances`. Based on the ***Process_Template*** `Instance` and ***Information*** `Instances` created in Step 1 and Step 5, create the ***GeneralDesign_Knowledge*** `Instances` and edit the generated `Instances`.

Step 7. Create ***Interface*** `Instances` and link each ***SPs_Entity*** `Instance` and ***Information*** `Instance`. Drag the yellow circle marked "Interface" and diamond marked "Information" from the left panel to the right canvas, and edit the generated `Instances`. The number of ***Interface*** `Instances` is determined by the number of dependencies among the ***SPs_Entity*** in a problem. Link the ***SPs_Entity*** `Instances` by specifying `Slots` "Process_1" and "Process_2" and "Lateral-Dependency", "Vertical-Dependency_Root" or "Vertical-Dependency_Leaf". Link the ***Information*** `Instances` and ***SPs_Entity*** `Instances` by specifying `Slots` "input" and "output". The ***Interface*** `Instances` implement the information flows between ***SPs_Entity*** `Instances` by instancing `Slots` "InformationFlow: 1–2" or "InformationFlow: 2–1". After these `Slots` are specified, the links are automatically shown on the canvas.

Step 8. Run the model to solve the problem with the DSIDES software and embed the results into the corresponding `Slots`. This is done with Java function calls in Protégé, with further details on the process described in [29].

- Design Scenarios of Process Templates

After the ontology development, it is essential to create and populate instances. In an ontological environment, most of the populated instances can be reused in future designs. While to successfully reuse these instances, various design scenarios for the creation of process template instances are defined, the main types of scenarios for process templates can be summarized as follows:

DS1: Design Process Template—this type of design scenario includes (a) creating an instance of process template and populating its slots, (b) adding a new system object instance, (c) adding the new necessary knowledge and related information for the design phase, (d) populating slots for a particular system instance based on existing knowledge and information, and (e) creating the phase support problem entities and linking them using instantiated interfaces.

DS2: Phase Process Template—this type of design scenario includes (a) creating an instance of process template and populating its slots, (b) creating the event support problem entities in the appropriate phase, (c) adding the necessary knowledge and related information for the design event, (d) enriching the existing design process templates based on the results of the phase, and (e) linking each of event entities using instantiated interfaces.

DS3: Event Process Template—this type of design scenario includes (a) creating an instance of a process template and populating its slots, (b) creating the decision/task support problem entities in the appropriate event, (c) adding the necessary knowledge and related information and its attributes for the design decision and task,

(d) enriching the existing phase process templates based on the results of the event, and (e) linking each of decision/task entities using instantiated interfaces.

DS4: Decision Process Template—this type of design scenario includes (a) creating an instance of process template and populating its slots, (b) creating the DSP template for solving an appropriate decision support problem, and populating its corresponding slots based on the current information, (c) creating the task support problem entities in the decision, (d) adding the necessary knowledge and related information and its attributes for the design tasks, (e) enriching the existing event process templates based on the results of the decision, and (f) linking each decision/task entity using instantiated interfaces.

DS5: Task Process Template—this type of design scenario includes (a) creating an instance of a process template and populating its slots, (b) adding the necessary knowledge and related information for the design task, (c) populating corresponding slots based on the existing information and knowledge, (d) enriching the existing event/decision process templates based on the results of the task, and (e) linking each of the task entities using instantiated interfaces.

5.4 Test Example: Design of Shell and Tube Heat Exchanger

In this section, we illustrate the validity of the PEI-X ontology through an example of a heat exchanger design.

5.4.1 Design of Shell and Tube Heat Exchanger for Thermal System

The small-scale power plant thermal system is widely used in farm irrigation. It produces electricity by using direct mechanical power or running small generators. The primary components of the small thermal system include a turbine to produce power, a pump to pressurize the flow to the turbine, a condenser, and a heat exchanger [32]. In the thermal system design, a decision-based approach is used to deal with design problems by formulating a DSP model with cycle efficiency issues [32], the choice of energy collector issues [33], and heat exchanger design issues [34]. In this section, our focus is on the meta-design process of the heat exchanger for a thermal system.

The heat exchanger is a heat transfer device transferring internal thermal energy between two or more available fluids at different temperatures [35]. Heat exchangers are classified into many types, each having its construction and performance features, which determine the corresponding application. The shell and tube type is the most commonly used heat exchanger, which has frequent applications in power generation,

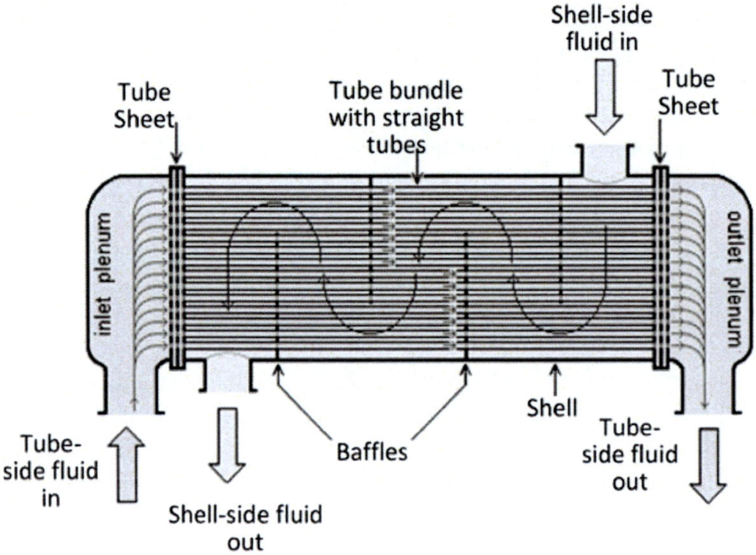

Fig. 5.5 Shell and tube heat exchanger (one pass tube-side) [28]

the chemical industry, and the process industry [32]. As shown in Fig. 5.5, the major components of a Shell and Tube Heat Exchanger (STHE) are a tube, baffles, shell, and tube sheets.

The design methodology of a heat exchanger is covered in detail by Hewitt and Barbosa [36], with the two most common heat exchanger design problems being sizing and rating. The sizing problem involves selecting exchanger construction type, tube and shell material, flow arrangement, and physical sizes of an exchanger to meet the specified heat transfer and pressure drop requirements. The rating problem involves performance evaluation, such as the transfer efficiency and pressure drop. The design of an STHE design process involves the following primary design considerations: (1) thermodynamic design to ensure the required performance of the heat exchanger and to satisfy the requirements of pressure drop for each stream, (2) mechanical design to provide the mechanical integrity required by the design codes and operating conditions, i.e., TEMA Standards, ANSI/API Standard 660, and (3) cost and manufacturing considerations. Most of these factors are interdependent and must be considered simultaneously to achieve the appropriate exchanger design.

5.4.2 Using DSPT Palette Entities for the Shell and Tube Heat Exchanger Design

As shown in Fig. 5.6, the top-level design process of STHE is partitioned into three major design phases, i.e., *Designing for Concept*, *Designing for Manufacture*, and

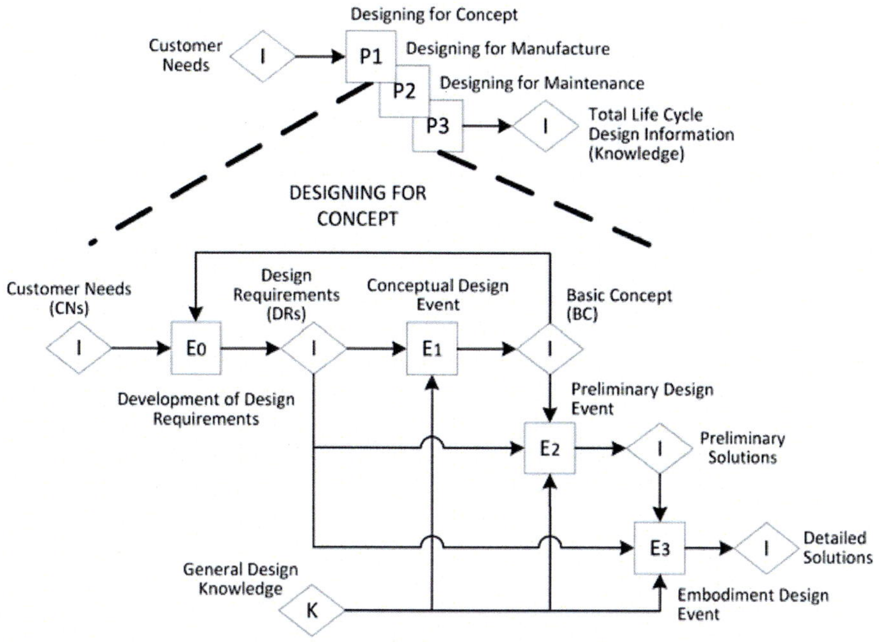

Fig. 5.6 A model of the designing for concept phase [28]

Designing for Maintenance, and the input and output information, i.e., *Customer Needs* and *Total Life Cycle Design Knowledge.* In the *Designing for Concept Phase*, designers convert information relevant to the customers' needs and engineering requirements for a product into specific knowledge used in designing for manufacture and maintenance, e.g., *Detailed Solution for STHE Design Specification.* In the process of designing for manufacturing, designers seek solutions ensuring the product can be manufactured cost-effectively.

In a computer environment, e.g., platform PDSIDES, the lower-level process models can be displayed via the action of clicking on and opening an icon. For example, we opened the Phase icon for *Designing for Concept* in Fig. 5.6, where the events constituting this phase are visible. The first event of this phase is the *Development of Design Requirements* (E_0) and results in a document. This document involving general design information forms the Conceptual Design Event (E1) input, which feeds forward a basic concept while initiating a feedback loop to E_0. The information relevant to *Basic Concept* and the *General Design Knowledge* provides the necessary inputs for the *Preliminary Design Event* (E_2) and *Embodiment Design Event* (E_3). E_2 provides a top-level specification, i.e., *Preliminary Solutions*, and E_3 provides a general specification, i.e., *Embodiment Solutions.*

Clicking on the icon of *Conceptual Design Event*, the detailed model of this event is given in Fig. 5.7. The primary goal of this event is to establish the *Basic Concept* (BC). Therefore, *Concept Alternatives*, e.g., Tubular Heat Exchanger and Plate Heat Exchanger, must be generated using *General Design Knowledge.* Then,

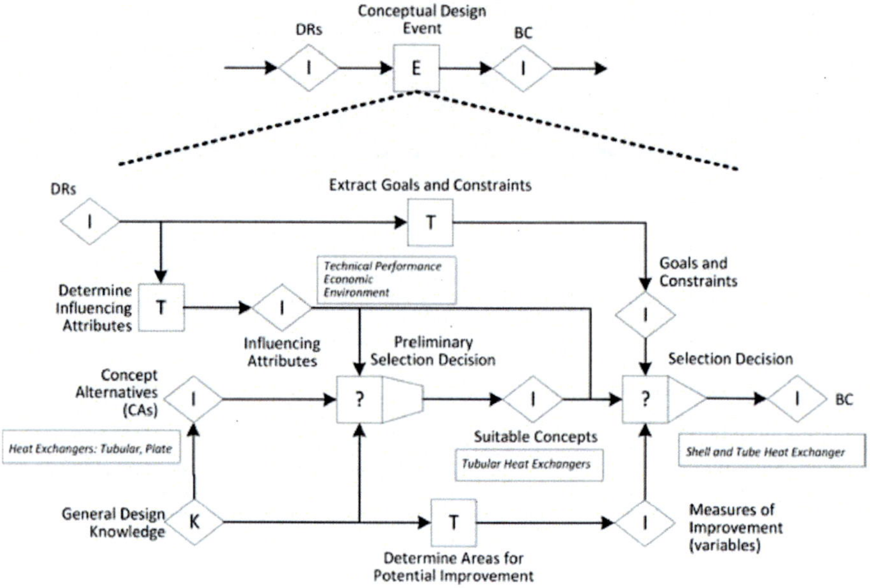

Fig. 5.7 A model of the conceptual design event [28]

a *Preliminary Selection Decision* is made to identify the *Suitable Concepts*, e.g., Coiled Tube, Double Pipe, Shell, and Tube, set for further development, and a *Selection Decision* is solved to identify the *Basic Concept* (e.g., Shell and Tube Heat Exchanger). *Influencing Attributes* based on the considerations of technical performance, economic and environment, etc., are needed for both the *Preliminary Selection Decision* and the *Selection Decision.* The *Goals and Constraints* extracted from the *Design Requirements* (DRs) are also needed to solve the problem of *Selection Decision.* The tasks of *Determine Influencing Attributes, Extract Goals and Constraints* from the *Design Requirements,* and the generation of concepts can be performed sequentially or concurrently.

Once the *Basic Concept* is known from the *Conceptual Design Event*, designers can start the *Preliminary Design Event*. A process model of the *Preliminary Design Event* is shown in Fig. 5.8. Here, the basic concept needs to be refined based on the necessary knowledge. The subsystems and their related information must be determined, e.g., *Material Concepts* and *Mechanical Concepts, Thermal Analysis.* Based on the information, an STHE Design Synthesis needs to be performed. Thus, material selection and dimensions for STHE have to be determined.

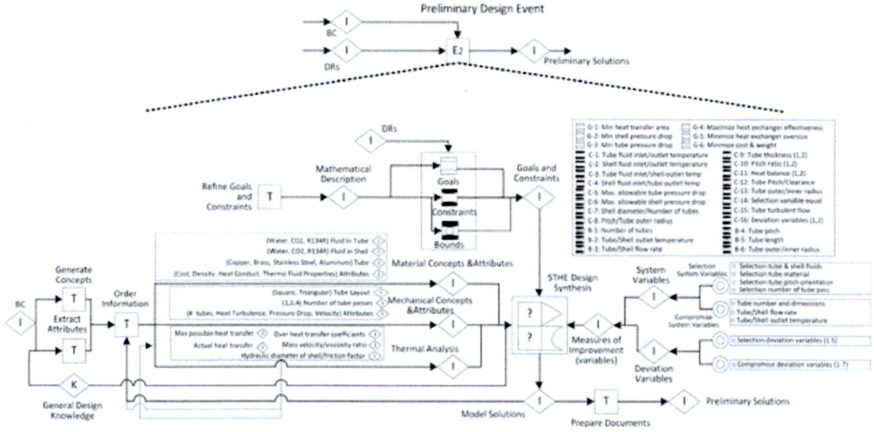

Fig. 5.8 A model of the preliminary design event [28]

5.4.3 *Design Scenarios for Shell and Tube Heat Exchanger Process Templates*

According to the process model for STHE design utilizing the DSPT Palette, the specified design scenarios of process templates will be illustrated by building a meta-design process hierarchies model via the PEI-X ontology.

DS1: STHE Design Process Template

The purpose of the design process template is to represent and document the information that characterizes the customers' needs into specific specification knowledge of a product object in the design timeline. For this example, the customers' needs are to design a heat exchanger for the small-scale power plant thermal systems, whose object is to achieve a minimum heat transfer area, minimum pressure drop, maximum effectiveness, and minimum cost. To meet the demands, the phase support problems for the designing involve three-phase instances in `Class` ***SPs_Entity***, i.e., P1- "Designing for Concept", P2- "Designing for Manufacture" and P3- "Designing for Maintenance", which are denoted with a blue box in the hierarchy graph, as shown in Fig. 5.9.

In Fig. 5.9, the interactions of information flow between phase entities are shown via the round yellow entities, i.e., "Interface-#1" and "Interface-#2". For example, the phase entities "P1" and "P2" are connected by the vertical type instance "Interface-#1", where the information flows "Design Solution of Concept Phase" and "Improvement for Concept Phase Solution" are defined, respectively.

The design process hierarchy is instantiated as a "STHE_Design" process template, presented in window "①" shown in Fig. 5.10. There are four plug-in instances of `Class` ***Process_Template***, with the details of the plug-in instance explained in Sect. 5.3.

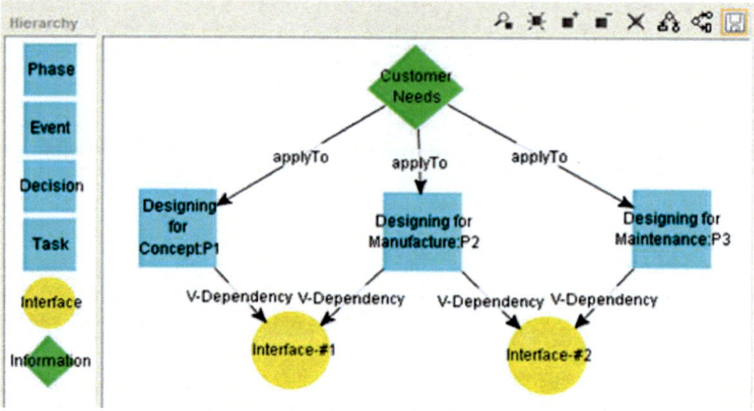

Fig. 5.9 Process hierarchies for design process template [28]

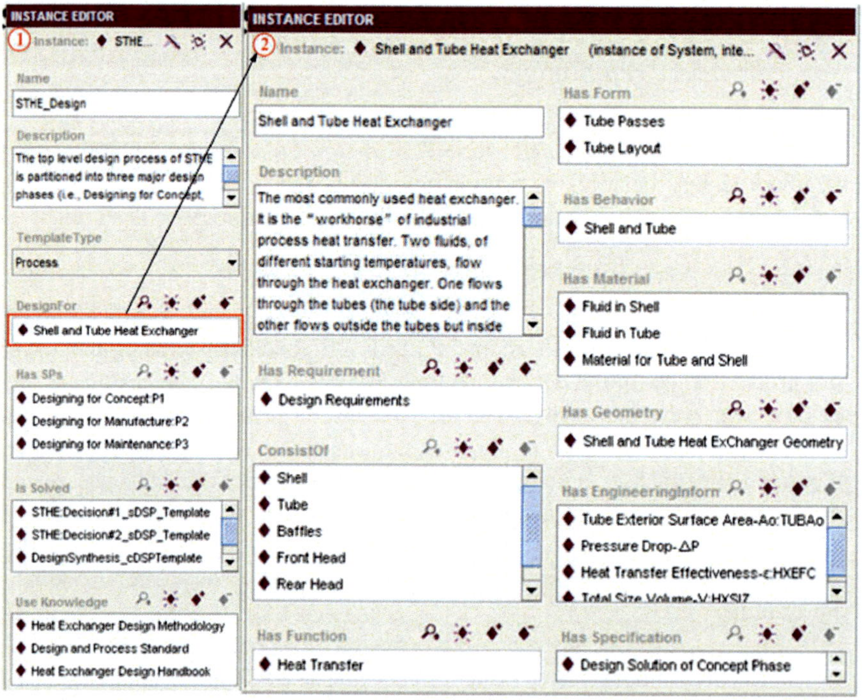

Fig. 5.10 Specification of the model for design process template [28]

As one of the primary instances, the `Class` ***Sys_Entity*** embedded in the `Instance` "STHE_Design" is presented in window "②". Based on the result of the concept phase process, the main specification information of the product has been instantiated and populated in the corresponding `Slots` of `Instance` "Shell and Tube Heat Exchanger". Such as the behavior alternatives of a heat exchanger are chosen as the `Instance` "Shell and Tube" type by the conceptual design event using a preliminary selection decision support problem template and selection decision support problem template. The engineering information of the `Instance` "Shell and Tube Heat Exchanger" is identified by the preliminary design event using a selection-compromise coupled decision support problem template, e.g., tube exterior surface area (A_o), pressure drop (ΔP), heat transfer effectiveness (ε).

By separating design activities and their interfaces, information related to the problem, product, and process in the STHE design enables designers to utilize existing knowledge, and the design processes can be composed and reconfigured easily. The representation and document of decision workflows for STHE design can also be achieved by designing meta-design process hierarchies. Due to the consistent structure of the process templates [37] and the consistent maintenance of DSP templates [29, 30], designers can create, update, and reuse decision workflows.

DS2: STHE Concept Phase Process Template

The purpose of the phase process template is to partition the design problem into an event support problem and adding the new necessary knowledge and related information for the design event. In the design of STHE and according to some general design knowledge that is embedded in the phase instance "Heat Exchanger Design Methodology", the phase process of the heat exchanger can be divided into four events: E0- "Development of Design Requirements", E1- "Conceptual Design Event", E2- "Preliminary Design Event", and E3- "Embodiment Design Event". Each event has its corresponding input and output information.

In Fig. 5.11, the phase process hierarchies are instantiated as the "STHE_ConceptPhase" process template, which is presented in window "①" of Fig. 5.12. There are four plug-in instances of `Class` ***SPs_Entity***, "Conceptual

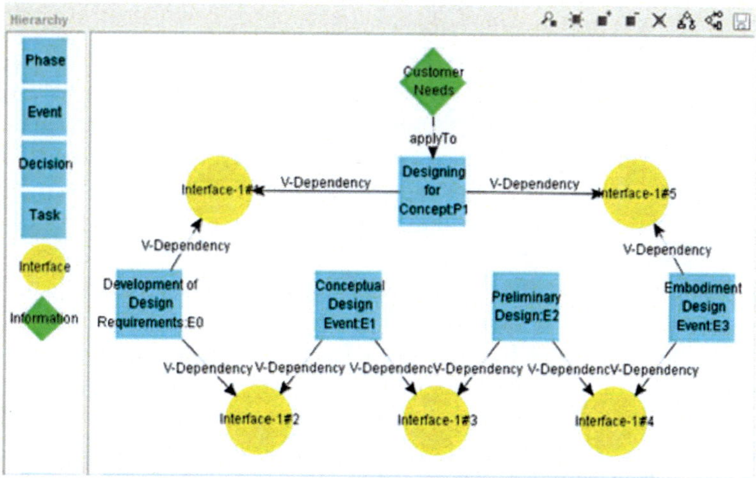

Fig. 5.11 Process hierarchies for phase process template [28]

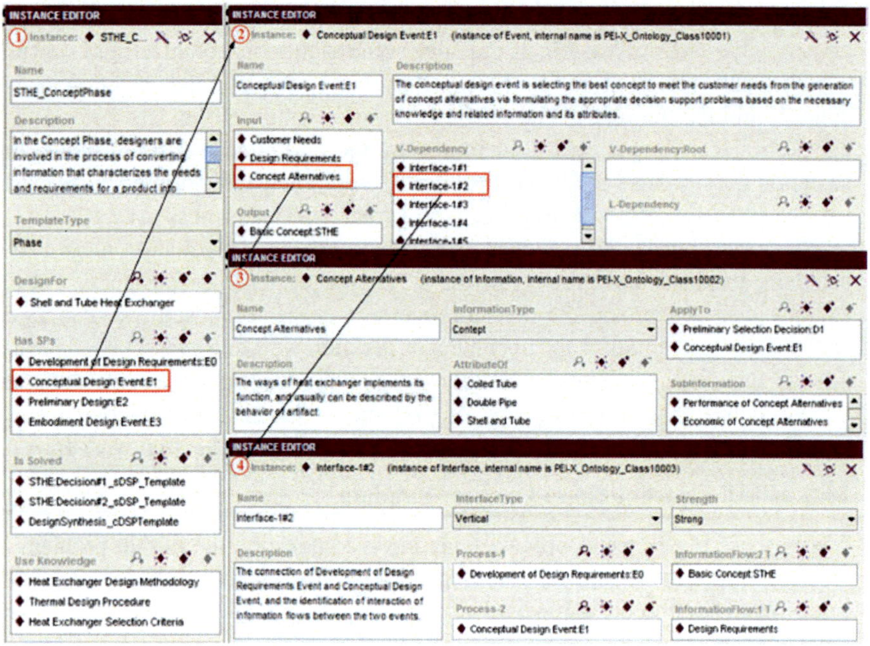

Fig. 5.12 Specification of the model for phase process template [28]

Design Event: E1" is presented in window "②" and the design object of "Conceptual Design Event, where E1" is selecting the best concept to meet the customer needs from the generation of concept alternatives via formulating the appropriate decision support problems based on the necessary knowledge and related information and its attributes.

In Fig. 5.12, the information instance of "Concept Alternatives" is presented in window "③", and three interfaces between the two events are defined as "Interface-1#2", "Interface-1#3", and "Interface-1#4". "Interface-1#2" is presented in window "④", which is the connection to "Development of Design Requirements Event" and "Conceptual Design Event", and the identification of interaction of information flows between the two events. "Interface-1#1" and "Interface-1#5" are used to represent the relationship between the root and child process nodes, i.e., the vertical dependency of parent/subsystem and information flows, where "Interface-1#5" also shows a feedback loop.

A significant amount of new necessary knowledge and related information for the design event is added in the design of phase processes, such as selecting the best concept in the process of E1, the "likely-to-succeed" concept alternatives need to be generated, which have been associated with the behavior of the `Class` ***System***. Various types of heat exchangers, e.g., coiled tube, double pipe, shell, and tube, will be populated in the behavior slots of the `Class` ***System***. Meanwhile, the knowledge of the E1 process, e.g., heat exchanger selection criteria, will be populated.

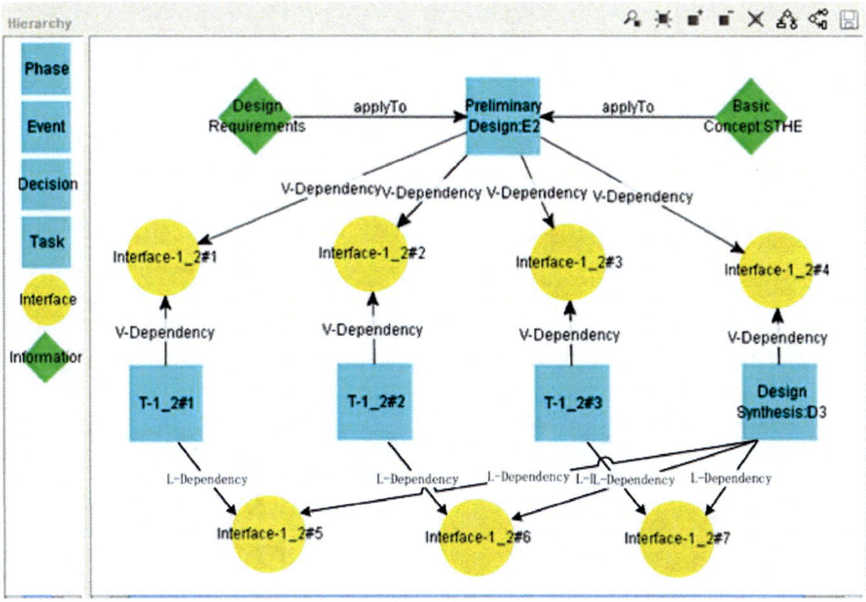

Fig. 5.13 Process hierarchies for preliminary design event process template [28]

DS3: STHE Event Process Template

The event process template is aimed to partition the design problem into a decision support problem and task support problem and plan their execution sequences. Moreover, the event process template adds the necessary knowledge, related information, and attributes for the design decision and task. In Fig. 5.13, we illustrate the process hierarchy of a preliminary design event for STHE design. This event explores a satisfactory solution as a feasible alternative, and the solution needs to meet specific design goals and constraints, which are accomplished by formulating and solving a selection-compromise coupled DSP. The event consists mainly of three tasks and one decision: "T1_2#1", "T1_2#2", "T1_2#3" and D3- "Design Synthesis", analyzed as follows:

(1) The objective of "T1_2#1" is to generate the design concepts and related attributes for STHE by the support of knowledge, such as, according to the knowledge of thermal design for STHE, the design of STHE needs three types of concepts, i.e., material concepts, mechanical concepts, and thermodynamic concepts (window "①" in Fig. 5.14). Each type of concept information has its corresponding sub-information, which will support the implementation of the decision-making activity, such as, the information about material concepts has three types of sub-information, i.e., "Fluid in Tube", "Fluid in Shell" and "Material for Tube", which are presented in window "②", "③", and "④", respectively. The related material attributes are presented in windows "⑤"

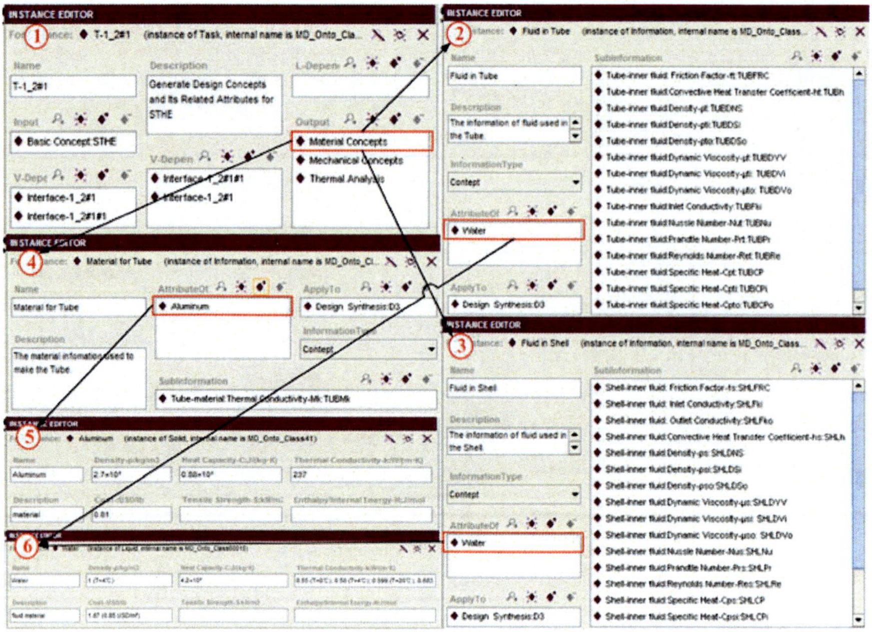

Fig. 5.14 Instance relationships for information of material concepts [28]

and "⑥", and the values of properties of the material can be populated based on the REFPROP [38] by using the Java function.

(2) "T1_2#2" is employed to refine the goals and constraints for the decision-making activity in the design synthesis. As shown in Fig. 4.8, the output of this task includes six goals, sixteen constraints, and six bounds. This information is used to formulate the DSP.

(3) The object of "T1_2#3" is extracting variables based on the results of "T1_2#1". Hence, the system variables are determined by the material and mechanical concepts, e.g., tube and shell fluid, tube material, tube pitch orientation, and the number of tubes. To formulate the DSP, the deviation variables also need to be defined in this task.

(4) The object of "Design Synthesis: D3" is executing the selection-compromise coupled DSP based on the information results from the tasks mentioned above. The information flows among them are represented via the "Interface-1_2#5", "Interface-1_2#6", and "Interface-1_2#7".

DS4: STHE Decision Process Template

The decision process template is primary in the design of meta-design process hierarchies. Here, the focus is on the order of information for formulating and solving the DSP, involving the identification of input information for the DSP formulation and the documentation of the result by the "Task: Post Solution", as shown in Fig. 5.15.

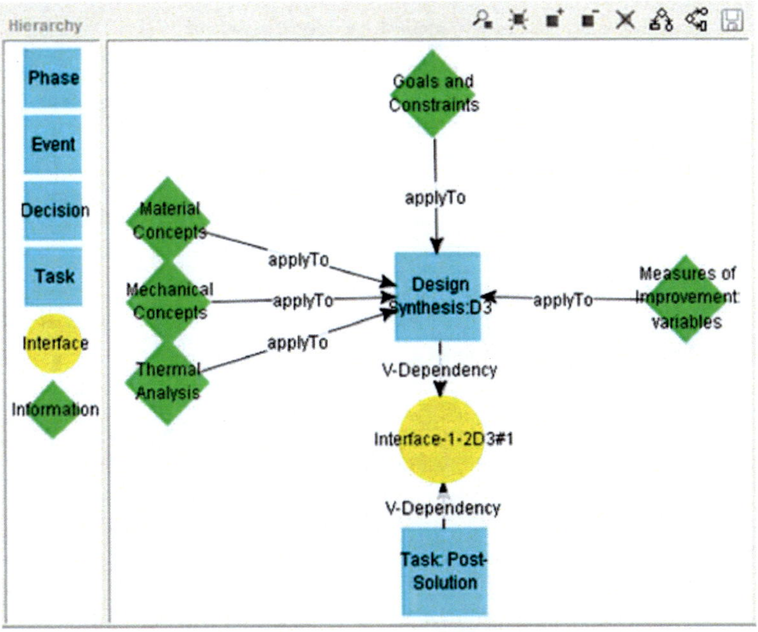

Fig. 5.15 Process hierarchies for design synthesis decision process template [28]

The order of information for the DSP formulation of design synthesis is shown in Fig. 5.16. The related information is populated into the `Slots` of DSP template, with the detailed description presented in [29].

DS5: STHE Task Process Template

The task process template is a fundamental activity in designing meta-design process hierarchies, focusing on the information order, namely, identifying input and output information. For example, in the thermal analysis process of "T1_2#1", the overheat transfer coefficient must be calculated to formulate the goal of heat transfer area (as shown in [34]). The formula of overall heat transfer coefficient (U) is

$$U = \frac{N_t}{\frac{1}{h_o} + \frac{r_o}{M_k}\ln\left(\frac{r_o}{r_i}\right) + \frac{r_o}{h_i r_i}}$$

where h_i and h_o are convective heat transfer coefficient for inner and outer fluid, respectively, *Nt* is the number of tubes, M_k the thermal conductivity of tube material, r_i and r_o are the tube inner and outer radius, respectively.

The relationship of the information within the previous equation is shown in Fig. 5.17, where the information flows between the calculated sub-task and parent task "T1_2#1" is represented by "Interface-1_2#1#1".

Given
Material concepts and their attributes with their relative importance
- Fluid Type in Tube – Water, CO_2, R134R
- Fluid Type in Shell – Water, CO_2, R134R
- Material Type of Tube - Copper, Brass, Stainless Steel, Aluminum
- Attributes - Cost, Density, Heat Conductivity, Thermodynamic Fluid Properties
- Relative Importance of Attributes – Express the preferences of the designers' knowledge

Mechanical concepts and their attributes
- Tube Layout - Square, Triangular
- Number of Tube Passes - 1, 2, 4
- Attributes - Number of tubes, Heat Turbulence, Pressure Drop, Velocity

Thermal analysis information
- Overall Heat Transfer Coefficients (U)
- Mass Velocity / Viscosity Ratio (G_t/ϕ_t)
- Hydraulic Diameter of Shell / Friction Factor (D_H/f_s)
- Max Possible Heat Transfer ($\dot{Q}_{max}$)
- Actual heat transfer ($\dot{Q}_{act}$)

The parameters of various concepts
Mathematical models of the goal description
Targets value for the goals, constraints, and bounds

Find
Selection system variables
X_s = 1 if material & mechanical concept is selected, X_s = 0 otherwise.
Selection Tube Fluid (STF_i): $X_1 = STF_i$ (i=1, 2, 3)
Selection Shell Fluid (SSF_j): $X_2 = STF_j$ (j=1, 2, 3)
Selection Tube Material (STM_k): $X_3 = STM_k$ (k=1, 2, 3, 4)
Selection Tube Pitch Orientation ($SIPCH_m$): $X_4 = SIPCH_m$ (m=1, 2)
Selection Number of Tube Pass ($SNTP_n$): $X_5 = SNTP_n$ (n=1, 2, 3)

Selection deviation variables
$d_i^-, d_i^+, d_j^-, d_j^+, d_k^-, d_k^+, d_m^-, d_m^+, d_n^-, d_i^+$

Compromise system variables
The Number of Tubes (NTUB) = N_t [-]
The Tube Length (TUBLEN) = L_T [m]
The Tube Outer Radius (TUBro) = r_o [m]
The Tube Inner Radius (TUBri) = r_i [m]
The Tube Clearance (TUBCLR) = C_t [m]
The Tube Outlet Temperature (Tto) = Tt_o [K]
The Shell Outlet Temperature (Tso) = Ts_o [K]
The Tube Mass Flow Rate (FLOWT) = $\dot{m}_t$ [kg/s]
The Shell Mass Flow Rate (FLOWS) = $\dot{m}_s$ [kg/s]

Compromise deviation variables
The under- and- over achievement of the heat transfer area goal, e_1^-, e_1^+
The under- and- over achievement of the shell pressure drop goal, e_2^-, e_2^+
The under- and- over achievement of the tube pressure drop goal, e_3^-, e_3^+
The under- and- over achievement of the heat exchanger effectiveness goal, e_4^-, e_4^+
The under- and- over achievement of the heat exchanger oversize goal, e_5^-, e_5^+
The under- and- over achievement of the heat exchanger cost goal, e_6^-, e_6^+
The under- and- over achievement of the heat exchanger weight goal, e_7^-, e_7^+

Satisfy
Selection system constraints (Only one alternative can be selected)
X_1: $\sum STF_i = 1$
X_2: $\sum SSF_j = 1$
X_3: $\sum STM_k = 1$
X_4: $\sum SIPCH_m = 1$
X_5: $\sum SNTPS_n = 1$

Selection goal constraints (Merit Function (MF), closest to 1)
Select Concept-1: $MF_i + d_i^- - d_i^+ = 1$ (i =1, 2, 3)
Select Concept-2: $MF_j + d_j^- - d_j^+ = 1$ (j =1, 2, 3)
Select Concept-3: $MF_k + d_k^- - d_k^+ = 1$ (k =1, 2, 3, 4)
Select Concept-4: $MF_m + d_m^- - d_m^+ = 1$ (m =1, 2)
Select Concept-5: $MF_n + d_n^- - d_n^+ = 1$ (n =1, 2)
Where, $MF_i = \sum_{i=1}^{n} I_j R_{ij}$ i=1,2,....,m
I_j = relative importance of jth attribute
R_{ij} = rating of alternative i for the attribute j
MF_i = value of merit function for alternative i
m = number of alternatives
n = number of attributes

Compromise system constraints
C-1: $Tt_i \geq Tt_o$
C-2: $Ts_i \leq Ts_o$
C-3: $Tt_i * 1.02 \geq Ts_o$

Fig. 5.16 DSP formulation for design synthesis process of STHE design [28]

C-4: $Ts_i \le Tt_o * 1.02$
C-5: $2.5 < P_T / r_o$
C-6: $P_T / r_o < 3$
C-7: $\Delta P_t \le$ Maximum Pressure Drop in Tube (TMXPD)
C-8: $\Delta P_s \le$ Maximum Pressure Drop in Shell (SMXPD)
C-9: $T_t \ge$ Minimum Tube Thickness (TMITH)
C-10: $T_t \le$ Maximum Tube Thickness (TMXTH)
C-11: $r_o \ge r_i$
C-12: $\dot{Q}_{in} \le \dot{Q}_{out}$
C-13: $\dot{Q}_{in} \ge \dot{Q}_{out}$ - (HLSMX $* \dot{Q}_{out}$)
C-14: $P_T \ge C_t + 2 * r_o$
C-15: $C_t > 0$
C-16: $Re_t \ge 4000$
C-17: $e_i^- * e_i^+ = 0$
C-18: $e_i^-, e_i^+ \ge 0$

Compromise goal constraints

G-1: Minimize heat transfer area: $\frac{A_0}{A_{0Target}} + e_1^- - e_1^+ = 1$

G-2: Minimize shell pressure drop: $\frac{\Delta P_s}{\Delta P_{sTarget}} + e_2^- - e_2^+ = 1$

G-3: Minimize tube pressure drop: $\frac{\Delta P_t}{\Delta P_{tTarget}} + e_3^- - e_3^+ = 1$

G-4: Maximize heat exchanger effectiveness: $\frac{\varepsilon}{\varepsilon_{Target}} + e_4^- - e_4^+ = 1$

G-5: Minimize heat exchanger oversize: $\frac{HXSIZ}{HXSIZ_{Target}} + e_5^- - e_5^+ = 1$

G-6: Minimize heat exchanger cost: $\frac{HXCOS}{HXCOS_{Target}} + e_6^- - e_6^+ = 1$

G-7: Minimize heat exchanger weight: $\frac{HXWEI}{HXWEI_{Target}} + e_7^- - e_7^+ = 1$

Bounds on system variables

B-1: Minimum value$\le N_t \le$ Maximum value
B-2: Minimum value $\le L_t \le$ Maximum value
B-3: Minimum value $\le r_o \le$ Maximum value
B-4: Minimum value $\le r_i \le$ Maximum value
B-5: Minimum value $\le C_t \le$ Maximum value
B-6: Minimum value $\le Tt_o \le$ Maximum value
B-7: Minimum value $\le Ts_o \le$ Maximum value
B-8: Minimum value $\le \dot{m}_s \le$ Maximum value
B-9: Minimum value $\le \dot{m}_t \le$ Maximum value

Minimize

The objection function (Lexicographically)

$$Z = \{P_1(e_1^-, e_1^+, e_2^-, e_2^+, \ldots, e_7^-, e_7^+) + P_2(d_i^-, d_i^+, d_j^-, d_j^+, d_k^-, d_k^+, d_m^-, d_m^+, d_n^-, d_n^+)\}$$

Where P_1 is the highest preference, P_2 corresponds to the selection-related deviation variables.

Fig. 5.16 (continued)

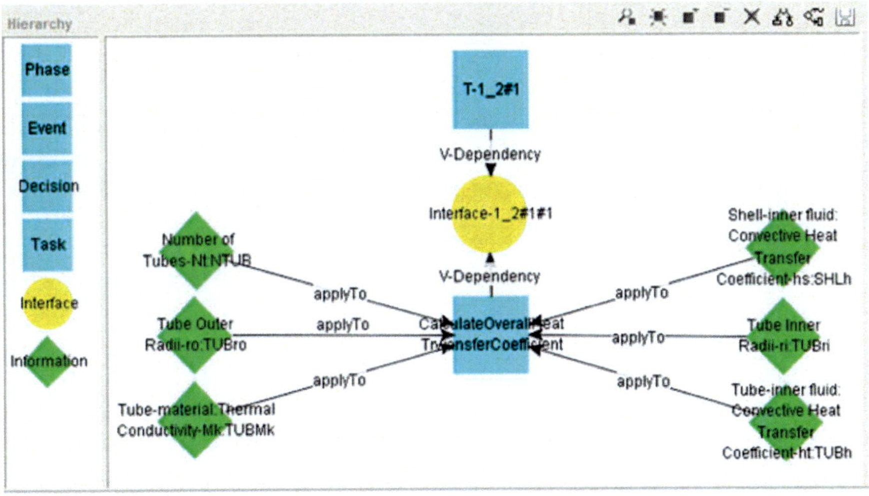

Fig. 5.17 Process Hierarchies for the Calculate Task Process Template [28]

5.5 Empirical Structural Validity

Empirical structural validation is to build confidence in the appropriateness of the test example problems chosen for illustrating and verifying the performance of the framework and methods. Using the design process template to design meta-design process hierarchies involves deciding how the process templates are instantiated, either as a top-down design decomposition or as a bottom-up design evolution. Hence, the decomposition approach of the process node needs to be considered during the designing of process templates. In [39], a method is presented for determining how the process node is designed as a top-down design decomposition node or a bottom-up design evolution node. In this section we focus on the following points: (1) The type of design scenarios for meta-design process hierarchies. In practical engineering design, different scenarios may be designed sequentially or concurrently based on o specific requirements. (2) Another critical focus is the purpose of various design scenarios. The scenario of the design process template concerns the product information, e.g., design specification, with the satisfaction of customers' needs and design requirements. The partition and planning of the design problems are the critical points in the design scenario of the phase and event process templates. In contrast, the design scenarios for decision and task process templates focus on formulating and solving a specific design problem. (3) Identifying the information interaction flows between the two process entities, represented by the interface instance.

5.6 Where We Are and What Comes Next?

In this chapter we develop an ontology for representing the knowledge related to meta-design of decision workflows, which serves as the foundation for providing knowledge support in designing of decision-based design processes on the PDSIDES platform. The next chapter develops ontologies for representing knowledge related to design space exploration and uncertainty management in designing decision workflows.

References

1. Noy, N. F., & McGuinness, D. L. (2001). Ontology development 101: A guide to creating your first ontology. Stanford knowledge systems laboratory technical report KSL-01–05 and Stanford medical informatics technical report SMI-2001–0880, Stanford, CA
2. Wang, W., De, S., Toenjes, R., Reetz, E., & Moessner, K. (2012). A comprehensive ontology for knowledge representation in the internet of things. In *2012 IEEE 11th International Conference on Proceedings of Trust, Security and Privacy in Computing and Communications (TrustCom)* (1793–1798). IEEE.
3. Lin, H.-K., Harding, J. A., & Shahbaz, M. (2004). Manufacturing system engineering ontology for semantic interoperability across extended project teams. *International Journal of Production Research, 42*(24), 5099–5118.

4. Chun, S., & Atluri, V. (2003). Ontology-based workflow change management for flexible eGovernment service delivery. In *Proceedings of the 2003 Annual National Conference on Digital Government Research* (pp. 1–4). Digital Government Society of North America.
5. Preuveneers, D., Van den Bergh, J., Wagelaar, D., Georges, A., Rigole, P., Clerckx, T., Berbers, Y., Coninx, K., Jonckers, V., & De Bosschere, K. (2004). Towards an extensible context ontology for ambient intelligence. In *Proceedings of European Symposium on Ambient Intelligence* (pp. 148–159). Springer.
6. Kitamura, Y., & Mizoguchi, R. (2004). Ontology-based systematization of functional knowledge. *Journal of Engineering Design, 15*(4), 327–351.
7. Grüninger, M. (2004). *Ontology of the process specification language*. Springer.
8. Grüninger, M., Atefi, K., & Fox, M. S. (2000). Ontologies to support process integration in enterprise engineering. *Computational & Mathematical Organization Theory, 6*(4), 381–394.
9. Lee, J., Gruninger, M., Jin, Y., Malone, T., Tate, A., & Yost, G. (2007). Process Interchange Format (PIF) for sharing ontologies. University of Edinburgh.
10. Polyak, S. T., & Tate, A. (2000). A Common Process Ontology for Process-Centred Organisations, *Dai Research Paper*, pp. 115–125.
11. Lin, Y., & Krogstie, J. (2010). Semantic annotation of process models for facilitating process knowledge management. *IGI Global.*
12. Thomas, O., & Michael, F. M. A. (2009). Semantic process modelling—design and implementation of an ontology-based representation of business processes. *Business & Information Systems Engineering, 51*(6), 506–518.
13. Sim, S. K., & Duffy, A. H. B. (2003). Towards an ontology of generic engineering design activities. *Research in Engineering Design, 14*(4), 200–223.
14. Kumar, P. P. (2008). *Design process modeling: Towards an ontology of engineering design activities*. Masters thesis, Clemson University.
15. Green, S., Southee, D., & Boult, J. (2014). Towards a design process ontology. *The Design Journal, 17*(4), 515–537.
16. Torga, M. Å., Andreasen, M. M., & Marjanoviä, D. (2010). The design ontology: foundation for the design knowledge exchange and management. *Journal of Engineering Design, 21*(4), 427–454.
17. Mistree, F., Bras, B., Smith, W. F., & Allen, J. K. (1995). Modelling design processes: A conceptual, decision-based perspective. *Journal of Engineering Design and Automation, 1*(4), 209–221.
18. Muster, D., & Mistree, F. (1988). The decision support problem technique in engineering design. *International Journal of Applied Engineering Education, 4*(1), 23–33.
19. Panchal, J. H., Fernández, M. G., Paredis, C. J., Allen, J. K., & Mistree, F. (2004). Designing design processes in product lifecycle management: Research issues and strategies. In *Proceedings of ASME 2004 International Design Engineering Technical Conferences and Computers and Information in Engineering Conference* (pp. 901–913). American Society of Mechanical Engineers.
20. Panchal, J., Fernández, M., Paredis, C., & Mistree, F. (2004). Reusable design processes via modular, executable, decision-centric templates. In *Proceedings of 10th AIAA/ISSMO Multidisciplinary Analysis and Optimization Conference* (p. 4601).
21. Mistree, F., Smith, W., Kamal, S., & Bras, B. (1991). Designing decisions: Axioms, models and marine applications. *Proceedings of Fourth International Marine Systems Design Conference* (pp. 1–24).
22. Bras, B., & Mistree, F. (1991). *Designing design processes in decision-based concurrent engineering*, No. 0148-7191. SAE Technical Paper.
23. Mistree, F., Smith, W., & Bras, B. (1993). A decision-based approach to concurrent design. *Concurrent Engineering*, 127–158. Springer.
24. Panchal, J. H., Fernandez, M. G., Paredis, C. J. J., Allen, J. K., & Mistree, F. (2009). A modular decision-centric approach for reusable design processes. *Concurrent Engineering, 17*(1), 5–19.

25. Graph Widget of Protégé. Stanford University. Retrieved February 1, 2016, from http://protegewiki.stanford.edu/wiki/Graph_Widget_Tutorial_OWL.
26. Ming, Z., Yan, Y., Wang, G., Panchal, J. H., Goh, C. H., Allen, J. K., & Mistree, F. (2018). Ontology-based representation of design decision hierarchies. *Journal of Computing and Information Science in Engineering, 18*(1), 011001.
27. Fenves, S., Foufou, S., Bock, C., Bouillon, N., & Sriram, R. (2004). CPM2: A revised core product model for representing design information. National Institute of Standards and Technology, NISTIR, 7185.
28. Wang, R., Wang, G., Yan, Y., Sabeghi, M., Ming, Z., Allen, J. K., & Mistree, F. (2019). **Ontology-based representation of meta-design in designing decision based workflows.** *Journal of Computing and Information Science in Engineering, 19*(march), 011003.
29. Ming, Z., Yan, Y., Wang, G., Panchal, J., Goh, C. H., Allen, J. K., & Mistree, F. (2015). Ontology-based executable design decision template representation and reuse. In *Proceedings of ASME 2015 International Design Engineering Technical Conferences and Computers and Information in Engineering Conference*, Paper Number DETC2015-46272.
30. Ming, Z., Wang, G., Yan, Y., Allen, J. K., & Mistree, F. (2017). An ontology for reusable and executable decision templates. *Journal of Computing and Information Science in Engineering, 17*(3), 031008.
31. Protégé 3.5 Release. Stanford University. Retrieved February 1, 2016, from http://protegewiki.stanford.edu/wiki/Protege_3.5_Release_Notes.
32. Smith, W. F., Milisavljevic, J., Sabeghi, M., Allen, J. K., & Mistree, F. (2015). The realization of engineered systems with considerations of complexity. *International Conference on Complex Systems Design & Management* (p. 55), November 12–14, 2014, Paris, France.
33. Bascaran, E., Mistree, F., & Bannerot, R. B. (1987). Compromise: An effective approach for solving multiobjective thermal design problems. *Engineering Optimization, 12*(3), 175–189.
34. Sabeghi, M., Smith, W., Allen, J. K., & Mistree, F. (2015). Solution space exploration in model-based realization of engineered systems. In *Proceedings of ASME 2015 International Design Engineering Technical Conferences and Computers and Information in Engineering Conference*, Paper Number DETC2015-46457.
35. Kaminski, D. A., & Jensen, M. K. (2005). *Introduction to thermal and fluids engineering.* Wiley.
36. Hewitt, G., & Barbosa, J. (2008). *Heat exchanger design handbook.* Begell House Connecticut.
37. Panchal, J. H., Gero Fernández, M., Paredis, C. J., Allen, J. K., & Mistree, F. (2009). A modular decision-centric approach for reusable design processes. *Concurrent Engineering, 17*(1), 5–19.
38. **Lemmon, E. W., & Huber, M. L. (2013). Implementation of pure fluid and natural gas standards: Reference fluid thermodynamic and transport properties database (REFPROP).** National Institute of Standards and Technology.
39. Fathianathan, M., & Panchal, J. H. (2009). Incorporating design outsourcing decisions within the design of collaborative design processes. *Computers in Industry, 60*(6), 392–402.

Chapter 6
Knowledge-Based Robust Design Space Exploration

Check for updates

Understanding and predicting process behavior in complex engineering systems design is important, primarily when incomplete and inaccurate models with unequal fidelity are employed. Indeed, this case requires designers to have the ability to make robust, modifiable, and flexible design decisions, especially in the early design stages. To address this need, in this chapter, we propose ontologies for robust design and a template-based ontological method that facilitates decision workflows, ensuring the proper design information combination meets various goals and requirements in a process chain. This is managed by utilizing (1) procedural knowledge—we define the arrangement of activities required to systematically explore the design space for a given uncertainty type, (2) declarative knowledge—we utilize the process templates to capture and document reusable information of a robust design and develop a frame-based ontology to formally represent tacit knowledge. The efficiency of our method is demonstrated using a hot rod rolling process design example that analyzes and synthesizes the microstructure-mechanical (rod module) and the processing-microstructure (cooling module) simulation models. A summary of this chapter is presented in Table 6.1. The mapping of the sections to the components (topics) discussed in this chapter is presented in Row 2 of Table 6.1.

6.1 Frame of Reference

Modeling a design process in a computational environment is treated as a template that can facilitate the executability and reusability of the domain-related information by separating the declarative knowledge, i.e., problem-specific information, from the procedural knowledge, i.e., process-specific information [1–3]. A template-based approach for modeling various design decisions and uniform representation of specific mathematical models is presented and validated in different application cases [1, 4, 5], and it provides modular support for human judgment in system design by means of the structured decision information content. An Extensible

© The Author(s), under exclusive license to Springer Nature Switzerland AG 2022
Z. Ming et al., *Architecting A Knowledge-Based Platform for Design Engineering 4.0*,
https://doi.org/10.1007/978-3-030-90521-7_6

Table 6.1 Summary interpretation of the problem investigated in this chapter

Elements	What?
Components	Ontology for systematic design space exploration (Sect. 6.2), ontology for designing robust decision workflows (Sect. 6.3)
Connectors	Design process and ontology
Form	**How?**
Component roles	Identify, capture, and represent the knowledge for robust design space exploration.
Properties	Requirements for developing the Design Space Exploration ontology and the Robust Design Process Ontology.
Relationship	The PEI-X diagram and the design space exploration procedure
Rationale	**Why?**
Motivation	To build a knowledge base for robust design space exploration.
Assumptions	There are four types of uncertainties in engineering design, the cDSP construct is used for formulating decisions in design
Interpretation	Understanding how robust design space exploration-related knowledge is represented using ontology

Markup Language (XML) procedure has also been used to create decision templates, providing a standard and easy-to-use tool to capture information [2]. Nevertheless, the consistency and inherent structure of the way in which the reusable information is used necessitates the formalization and representation of knowledge in the decision-making process. As specifications of conceptualization, ontologies afford standard terms or vocabulary, including their relationships, enabling formal representation of domain-specific knowledge [6], and facilitating the creation and reuse of decision templates.

Ontologies are used in several application domains ranging from knowledge management, knowledge exchange, to semantic retrieval and information integration [7]. The broad application is due to its unique expressiveness, semantic interoperability, intelligent behavior, and flexibility [8]. Indeed, ontologies can capture and archive design information in a generic way and encapsulate the concepts related to an area that is ultimately shared and reused between teams and software agents in a distributed design environment [9]. For example, Rockwell et al. [10] suggest a Decision Support Ontology (DSO) to support collaborative designs during information exchange in decision-making processes. Noor et al. [11] highlight ontology's beneficial value for activity identification under uncertainty.

In the semantic community, there is continuing research on extending the ontologies' ability to formally represent uncertain knowledge and support reasoning without accurate data [12–14]. Conceptually, current methods augment and supplement the ontology modeling language—Web Ontology Language (OWL) with uncertainty and annotate it by employing Bayesian networks, i.e., probabilistic networks capable of expressing and assessing probabilistic knowledge [12, 15]. For instance, during a product family design, a feature preference probability models a customer's preference uncertainty toward specific product features, exploiting customer preference survey data. The customer's preference uncertainty impact and propagation are quantitatively evaluated via the ontology-based Bayesian network [16]. Furthermore, Costa et al. [17] suggest an ontology reference model as part of the Evaluation of Technologies for the Uncertainty Representation Working Group (ETURWG), belonging to the uncertainty representation and reasoning evaluation framework (URREF). Finally, the PR-OWL (Probabilistic Web Ontology Language) is a Bayesian ontology language utilized as a supporting tool for this type of application [14].

However, some practical limitations have also been realized with regards to industrial implementation, especially in terms of robust design for model-based complex engineered systems, the attention paid to representing and capturing process-related knowledge that refers to uncertainty management is still inadequate. Sim et al. [18] propose an ontology appropriate for generic engineering design purposes that integrates the uncertainty and ambiguity into the activity of defining. Singh et al. [19] suggest a decision support ontology that enhances a supply chain's resilience utilizing the uncertainty of disruption. In the context of DSPT, most previous work related to decision-making knowledge representation has been focused on the information under the case of certainty in design, nevertheless, uncertainty is a more commonly encountered during the engineering design process.

6.2 Ontology-Based Representation of Systematic Design Space Exploration

In this section, we develop an ontology to represent the Design Space Exploration (DSE) template. First, we identify the requirements for modeling the knowledge related to DSE. Second, we discuss the structure of a DSE template for addressing the requirements. Third, we develop an ontology based on the structure of the DSE template. Finally, we use a hot rod rolling design example to test the ontology's utility.

6.2.1 Requirements for Design Space Exploration

In model-based engineered systems, managing the uncertainty and complexity during the DSE processes is important. Kang et al. [20] highlight that a DSE framework is effective if it involves: (1) an appropriate representation for the design space, (2) an effective exploration method, and (3) an analysis process utilizing machine-assisted methods. Chong and Chen [21] suggest an architecture for guiding designers in the DSE processes within the conceptual design space. In our work, the exploration process is extended to the subsequent design stages, where the design space can be quantified. To further ensure the validity of the design, we identify the following requirements:

- **Support Decision—Centric Robust Design**

  Decision-Based Design (DBD) bridges the gap between a model and a physical world [22], emphasizing the crucial role of human designers during the decision-making process in a computer design environment. It is widely accepted that a design is considered a decision-making procedure involving a set of rational decisions satisfying the designer's preferences [23, 24]. Therefore, to achieve robust decision-making under the conditions of complexity and uncertainty, human designers typically exploit a specific set of methods that helps them to determine potential robust strategies. As one embodiment of DBD, DSP provides domain-specific mathematical models constructed as structured templates that formulate a suitable design space representation.
- **Support Understanding and Predicting of Process Behavior**

  Supporting the various decision-making requirements necessitates the design space exploration processes to afford various aggregation levels. These levels include analyzing, evaluating, synthesizing, and defining (each task to be performed) at several levels of detail employing techniques that guide a task sequence from one abstraction level to the next lower level. For this purpose, Computer-Aided Engineering (CAE) tools enhance efficiency and facilitate task accomplishment. Therefore, considering realizing model-based engineered systems [25, 26], applying methods, processes, and tools to explore the design space requires an environment integrating the related information and affording enhanced communications to facilitate human designers in understanding and predicting the process behavior in DSE.
- **Interaction and Visualization Support**

  Realizing model-based engineered systems in a computer environment is impossible without the information flows facilitating interaction among models. Given the complex features of engineered systems, the design process hierarchy should manage and organize the information flows to facilitate horizontal and vertical integration. Thus, integrating information flows across several design process dimensions and stages is essential. Additionally, visualization is also crucial in supporting a robust decision-making process in the design space.

6.2.2 Design Space Exploration Procedure

In this section, we propose a design space exploration procedure that supports robust decision-centric designs by identifying design alternatives and generating *satisficing* solutions for the examined design problem. The suggested procedure is inspired by RCEM [27] and CEF [28]. The DSE frame is a logical sequence of activities achieving a particular objective (Fig. 6.1).

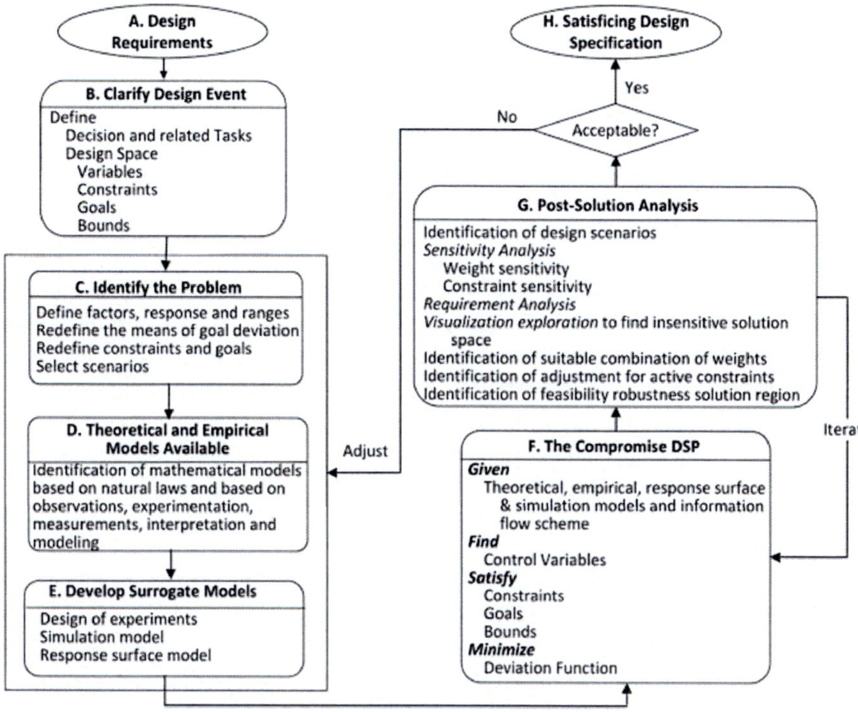

Fig. 6.1 Design Space Exploration Procedure [29]

Step 0: Data Input/Output—*Input A and Output H*

The DSE procedure starts with the designer setting the requirements by either providing data entry from a static problem statement or dynamic data, e.g., sensor data. The DSE ends by determining the design solution regions or points, which satisfy the requirements set and support the designer in making comprehensive decisions. Determining the design requires the consideration of stakeholders' conflicting desires. This is because to effectively deploy a product, both the consumers' and producers' needs should be incorporated in the decision-making process which facilitates the conceptualization of design alternatives and constraints [30].

Step 1: Pre-Process—*Processor B*

Through meta-design [31], partitioning a problem and planning the decisions processes employing generic discipline-independent modeling are presented in the DSPT. For example, a PEI-X (Phase-Event-Information - X) diagram models the design processes from an event-based time viewpoint. To ensure whether appropriate *Support Problems* are available to be solved via computer-based design and analysis of design space, the problem's complexity be further refined, i.e., clarify the design by setting the decisions and relevant tasks. The information associated with the design space, i.e., bounds, goals, constraints, and variables, is obtained from several sources to afford to model a specific problem.

Step 2: Problem Modeling—*Processor C, D, and E*

Determining the initial design space and providing a combination of design information as inputs for the cDSP model requires the designer to model the problem using mathematical formulations by performing three task processes (*Processor C, D,* and *E*). The first process involves identifying essential design parameters of the problem. These parameters are classified as control factors, **x,** that a designer controls, noise factors, **z,** that a designer cannot control, and responses, **y,** that measure the goal's performance. Additionally, the range of these parameters is also identified. Then, the functional relationship f between factors and responses, $\mathbf{y} = f(\mathbf{x})$, is defined. *Processor C* exploits the available empirical and theoretical mathematical models relying on existing natural laws or experimental/modeling knowledge. If some models are unavailable or it is required to develop models to reduce the problem size, designers have to develop surrogate/reduced-order models for the problem. To address these concerns, [32] statistical techniques are needed, e.g., statistical design of experiments and response surface methods, and a meta-model (model of the model) is developed by approximating the computer analysis codes to obtain an insight into the functional relationship between **y** and **x**. In Fig. 6.2, we illustrate a generic response surface modeling procedure that generates the prediction function, $g(\mathrm{x})$, approximating the true response surface function $f(\mathrm{x})$ using support tools and base steps.

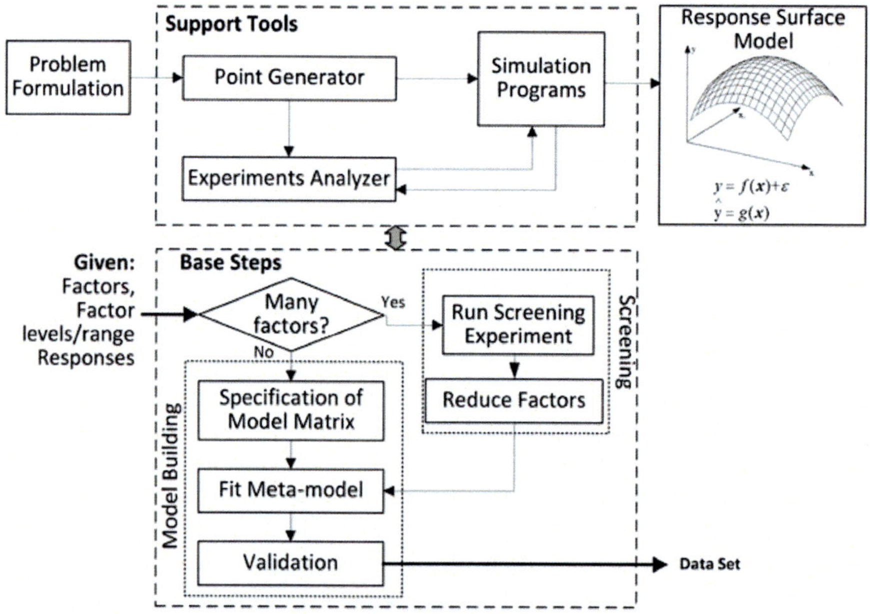

Fig. 6.2 Generic procedure for response surface modeling [29]

In the development processes of surrogate models, some candidate parameters and boundaries (i.e., factors, responses, and ranges) are defined as the inputs based on the existing knowledge. They characterize the design space for the identified problem as well as factor levels used in simulations for Design of Experiments (DoE). The *Simulation Program* as a "slot" for inserting Finite Element Analysis (FEA) and other simulation programs is used to run the experiments designed. Dataset generation for the response surface models involves two stages of sequential experimentation for building computer analysis approximations: screening and model building. For more details, the reader is referred to [33]. Finally, the *Point Generator* and *Experiments Analyzer* are used to design and evaluate the critical experiments and their results [27].

Step 3: Compromise DSP—*Processor F*

DSE's core step is the *compromise Decision Support Problem* (cDSP) that synthesizes information to design with multiple goals under uncertainty [34]. From the specific problem, the generated design information combination is input to *Processor F* that can handle multiple objectives, bounds, and constraints, minimize the deviation function and determine the design variable values satisfying the conflicting goals. Selecting the deviation function type between the Preemptive Formulation $Z = [f_1(d_i^-, d_i^+), \ldots, f_k(d_i^-, d_i^+)]$ and Archimedean Formulation $Z = \sum W_i(d_i^- + d_i^+)$, depends on whether a designer has sufficient information and knowledge to indicate the priority of the different objectives. Different design scenarios, defined by the design preference, P_i, associated with the design goals, G_i, and their weights, Wi, are established to explore the solution space through several executions of the cDSP model. To solve the cDSP, the computational tool DSIDES is used, which incorporates an Adaptive Linear Programming (ALP) algorithm [35] and a user-specified input file consisting of a data file that defines the design space size and FORTRAN routines to monitor the solution process.

Step 4: Post-Solution Analysis—*Processor G*

The notion of a multi-objective approach based on the cDSP formulation originates in an understanding of the problem defined by different performance criteria, which is appropriate in design under various uncertainties because it offers a "satisficing range" solution rather than an "optimizing point" solution. Rather than determining the optimum single-point solution (optimization philosophy), the satisficing philosophy pursues flexibility and modifiability during each design solution phase, especially in the early design phases. By defining the deviation functions, the designer can design the weight combination during post-solution analysis that guarantees a "satisficing range" of solutions and trade-off the multiple conflicting objectives. In Fig. 6.3, the desired solution is implemented by utilizing the design preferences to analyze the sensitivity of the weights of the goals. Different design scenarios are established and grouped based on the designer's requirements in the "scenarios experiments". The aim is to exercise these scenarios and explore the design space. The results are then visualized and analyzed through ternary plots and comparison charts to bring insight for decision-makers.

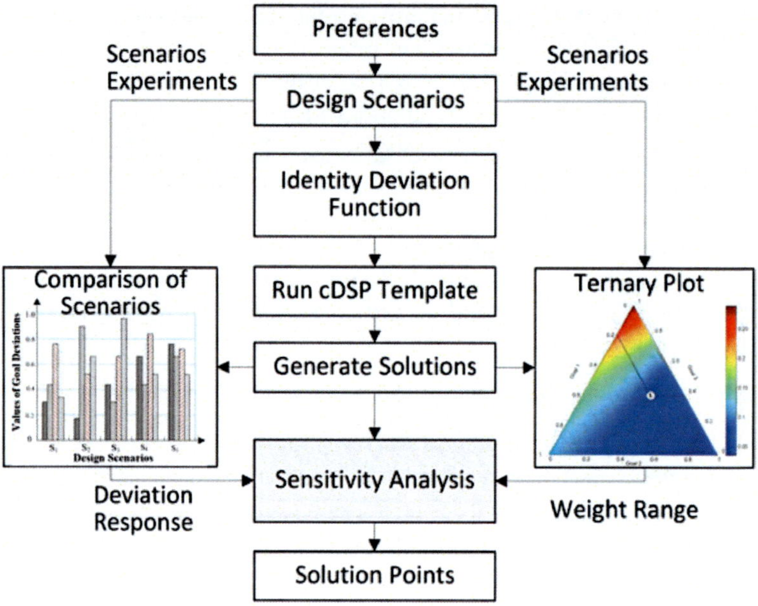

Fig. 6.3 Procedure for design preference exploration [29]

Regarding the comparison chart, the goal deviation change trend in the various design scenarios is a graphics display. Considering the ternary plot, the values inside the color contours represent the deviation linked with each system goal or the actual goal values per scenario. The color bar next to the triangles refers to the value changes. Based on the sensitivity analysis, several solution points satisfying the multiple design objectives from the feasible design space are identified and suggested as the design values or operating set points.

In the exploration processes, the designer obtains the deviation response in various design scenarios and determines a "satisficing range" solution meeting all the system goals by superimposing the plots into one ternary plot. If a region meets all goals simultaneously, the designer can alter the weight range to satisfy all the goals from the superimposed plot and then determine the solution values that include the goals and system variables. Another case is the inexistence of common regions [36]. In such a situation, it is essential to modify the target value of system goals assigned in the cDSP to lower the deviations and thereby enhance the overlap possible, or even reformulate the constraints/goals to adjust the feasible design space. Both cases are further analyzed in the following sections. Once the weight sensitivity analysis is completed, solution points belonging to the satisficing range are recommended to the designers, who have to trade-off the conflicting goals and choose a single solution point as input to the next stage according to their empirical knowledge and preference.

6.2.3 Design Space Adjustment

Considering the interdependencies among the different design events within a design process, the design space exploration processes must afford robustness and modifiability to mitigate the design errors due to other design stages, e.g., processing error. The constraints/goals of the cDSP model are functions of the system variables, namely, f(xi) and g(xi), which can get response according to the minimization of deviation variables under different design scenarios. Therefore, the design variables, goals, and constraints must be analyzed concerning feasibility robustness. In this section, we consider four possible scenarios in the design space (Fig. 6.4) that the designer can explore to identify a common satisficing region based on the design space requirements of the problem.

In Fig. 6.4, Scenario I, II, and III adjust target values associated with goals or ranges of the variables, goals, and constraints in the initial design space, respectively. In practice, the modifications are generally based on the designer's empirical knowledge and the corresponding comparison of the initial design results. Hence, a

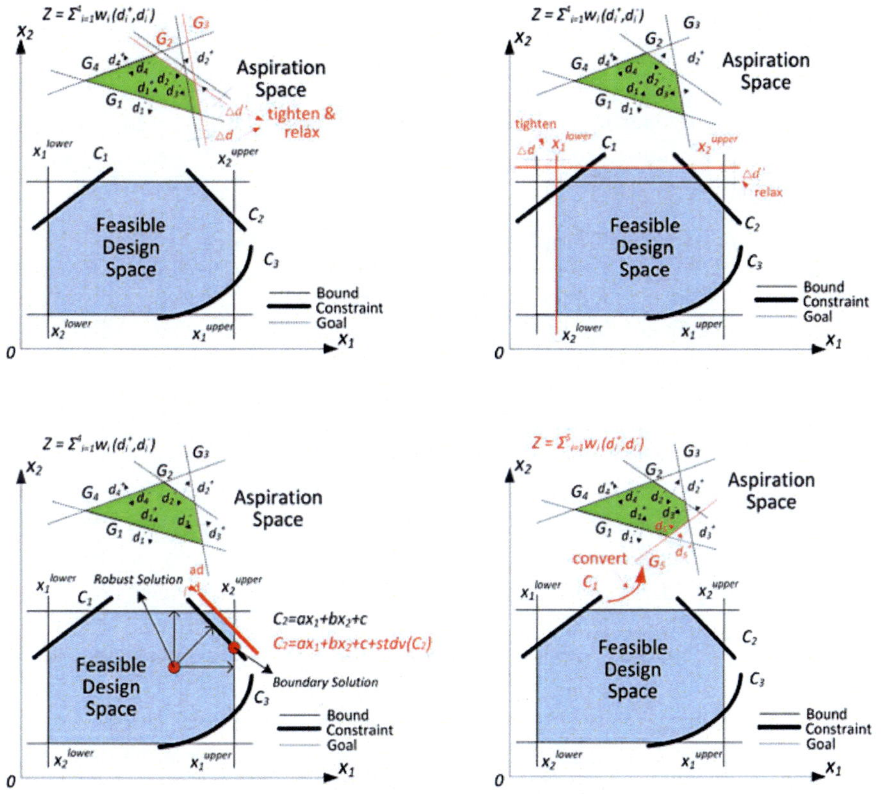

Fig. 6.4 Four possible scenarios for the design space adjustment [29]

detailed response analysis can increase the designer's confidence during decision-making. Regarding Scenario III, the additional design space capacity that depends on the constraints is determined by identifying the active constraints adjustments [37], reducing the risk of boundary solution with zero tolerance to become infeasible during variations. Hence, analyzing the constraint sensitivity affords the determination of the constraints that require modification by adding extra capacity. Regarding Scenario IV, in addition to the system goals, the designer considers new constraints or system variables requirements during his/her decision. Incorporating these "additional requirements" changes the design space, allowing the designer to make a confident design decision. Scenario IV that occurs in the design process of a multi-staged steel manufacturing procedure [28] is analyzed in Sect. 6.2.5. The appearance of those four scenarios depends on the specific design problem and the settings of the initial design space.

6.2.4 Ontology for Process of Design Space Exploration

We further satisfy the DSE requirements by developing a frame-based ontology for the DSE process template that supports information management of information reusability and enhances a designer's understanding of the process behavior. In this section, we define the slots and classes that establish a frame-based ontology and present the exploration processes that exploit the ontology by preserving the DSE process template model.

(1) **Modular Process Template for Design Space Exploration**

Modular-based design methods within a computational environment improve design flexibility and enhance design efficiency. Thus, we develop a modular-based process template model appropriate for design space exploration that manages reusability and executability. The DSE process template primarily includes three sub-templates: *Problem Model* (PM), *compromise Decision Support Problem* (cDSP), and *Post-Solution Analysis* (PSA). The PM sub-template involves two modules: *Theoretical and Empirical Model* and *Surrogate Model*, while PSA involves five modules: *Weight Sensitivity Analysis* (WSA), *Constraint Sensitivity Analysis* (CSA), *Additional Requirement Analysis* (ARA), *SSE_Experiment* (Solution Space Exploration Experiment), and *Deviation Response*. The cDSP's template modules are presented in detail in [4].

Figure 6.5 is the DSE process template similar to a printed board assembly that has some electronic components. The elements (modules), such as the empirical and theoretical models, and the deviation response, are denoted by "chips", while the process of Sect. 6.2.2 is the "breadboard". Given the modular structure, the DSE procedure template involves three reusability scenarios:

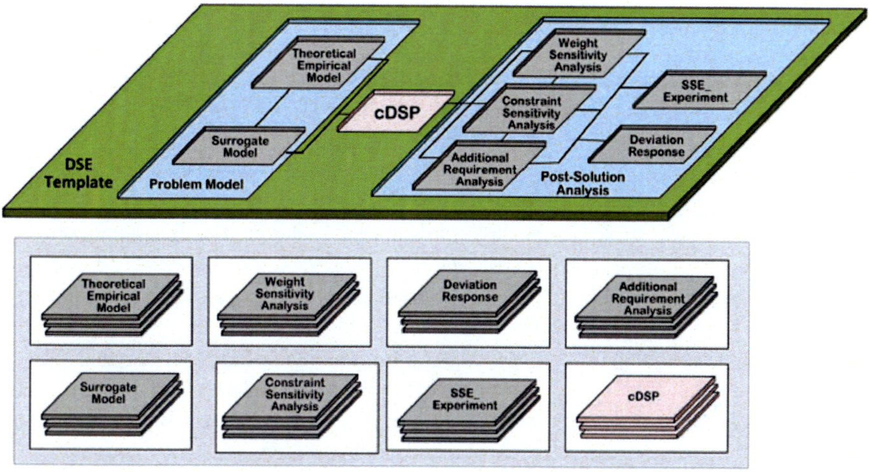

Fig. 6.5 The DSE process template [29]

(1) Reusing the "breadboard". The design space exploration process corresponding to the "breadboard" is reused by populating certain information on the board.
(2) Reusing the "chips". Specific information, such as the *Surrogate Model*, referring to the "chips," is reused in any problem variation for the exploration process template.
(3) Reusing the assembly. An instantiated DSE process template with certain information linked to the "chips" is reused, with some "chips", e.g., *SEE_Experiment*, being modified and others unchanged.

The modular DSE process template allows capturing and reusing the DSE's information, increasing the designer's decision-making confidence and providing insight to make comprehensive decisions, especially in the early design stages.

(2) Class and Slot Definition

The main ontology structure in the DSE process template comprises the "chips" embedded in the "breadboard". The concepts within the DSE process template are defined as `Classes`, such as *PSA_Template, PM_Template*, and *DSE_Template,* while additional associated `Classes`, such as *Factor, Response,* and *ResponseSurface,* capture DSE's reusability information that also increases the integrity and the semantic richness of the DSE process template ontology. Table 6.2 presents the detailed definitions of `Classes.`

`Slots` are used to capture the semantic relationships between `Classes`. Two types of `Slots` exist: data and objects. The former links classes to end data, e.g., weightRange links the WS_Analysis to capture a weight range value, while the latter type links classes to themselves or other classes, e.g., *hasWSA* links *PSA_Template* to *WS_Analysis*. Tables 6.2 and 6.3 define the ontology's data and object slots based on the exploration processes and the DSE process template structure. Slots that reuse other ontologies, i.e., *image, value,* and *name,* will not be described here (Table 6.4).

Table 6.2 DSE process template ontology classes [29]

Class	Definition
TernaryPlot	A class representing the visualizing information of the desired and sensitive regions of the solution space
SolutionPoint	A class representing a point's value in the specific satisficing range solution
Preference	A class representing the value of preference corresponding to the associated system goal in a specific design scenario
FactorLevel	A class representing the value of a factor level identified by the designers
FactorValue	A class representing the value of a specific factor corresponding to the associated factor level
DesignScenario	A class representing a set of preference values corresponding to the associated design weight
VariableResponse	A class representing the achieved value of the associated system variable in a specific design scenario
ConstraintResponse	A class representing the achieved value of the associated constraint in a specific design scenario, including "*Active Constraint*" and "Inactivate Constraint"
GoalDesponse	A class representing the achieved value of the associated system goal in a specific design scenario
GoalWeight	A class representing the designers' interest in the associated system goal
Factor	A class representing input variables corresponding to a specific process
Response	A class representing a mathematical model for performance measurement
ResponseSurface	A module integrating all the related information of a surrogate model using the response surface methodology
DeviationResponse	A module representing a set of goal deviations corresponding to the associated design scenario
SSE_Experiment	A module representing a set of design scenarios corresponding to the associated goal weight
AR_Analysis	A module integrating all the related information for the additional requirement analysis that defines a typical range solution
CS_Analysis	A module integrating all the related information of the constraint analysis to define extra capacity for the design space
WS_Analysis	A module integrating all the related information of weight analysis to calculate a range solution satisficing all system goals
SurrogateModel	A module integrating all the related information of the surrogate model and the experimental design

(continued)

Table 6.2 (continued)

Class	Definition
TheoreticalEmpiricalModel	A module integrating all the related information of a mathematical model for the initial design space
PSA_Template	A sub-template integrating all the associated modules and representing the information structure of solution space exploration
PM_Template	A sub-template integrating all the associated modules and representing the information structure for a specific problem
DSE_Template	A formulation integrating all the associated template modules and representing the DSE's process information structure

Table 6.3 DSE process template ontology data slots [29]

Class	Definition	Type
validationRSM	The response surface model verification results	String
typesOfFittingModel	The fitting model types representing a regression meta-model	Symbol
modelMatrix	The model matrix (path) representing the treatment combinations corresponding to the DoE types	String
simulationPrograms	The code execution (path) employed to run the simulation programs of the designed experiments	String
weightRange	The weight's range value for an associated goal satisfying all the system goals	Interval
deviationValue	A set of response values acting as a normalized treatment to generate the ternary plot	Float
acceptableValue	A value of the minimum requirements target that can be accepted or approved	Float
preferenceValue	A set of preference values for a specific design scenario and experiment of the solution space exploration	Float
achievedValue	A value achieved in response to the result of minimizing the deviation function	Float
extraCapactiy	A standard deviation value added to the active constraints with zero or limited capacity	Float
resluts_of_SSE	A set of values (system variables and goals) employed for solution points satisfying all the design requirements and goals	Float
dataPoint	A set of goal deviation values associated with a specific system goal that it employed to generate the ternary plot	Float
factorVaule	The value of a specific factor corresponding to the associated factor level that is employed in the DoE simulations	Float
lowest_SSE	The value of the lowest sum of squares error (highest R^2) used to fit the response regression model	Float

Table 6.4 DSE process template ontology object slots [29]

Class	Definition	Type
preferenceValue	Determines the *Preference* occurrence of the *DesignScenario*	Instance
toScenario	Determines the *DesignScenario* occurrence	Instance
associatedWeight	Determines the *GoalWeight* occurrence of the *Preference*	Instance
associatedConstraint	Determines the *Constraint* occurrence of the *AR_Analysis* and *ConstraintResponse*	Instance
associatedGoal	Determines the *Goal* occurrence of the *TernaryPlot, GoalDeviation, GoalWeight,* and *SolutionPoint*	Instance
associatedVariable	Determines the *Variable* occurrence of the *AR_Analysis* and *SolutionPoint*	Instance
constraintResponse	Determines the *ConstraintResponse* occurrence of the *CS_Analysis*	Instance
hasARA	Determines the *AR_Analysis* occurrence of the *PSA_Template*	Instance
hasCSA	Determines the *CS_Analysis* occurrence of the *PSA_Template*	Instance
hasWSA	Determines the *WS_Analysis* occurrence of the *PSA_Template*	Instance
toFactorLevel	Determines the *FactorLevel* occurrence of the *FactorValue*	Instance
associatedFactor	Determines the *Factor* occurrence of the *FactorValue*	Instance
functionOf	Determines the *Factor* occurrence of the *Response*	Instance
hasResponse	Determines the *Response* occurrence of the *ResponseSurface*	Instance
hasFactor	Determines the *Factor* occurrence of the *ResponseSurface*	Instance
hasTEM	Determines the *TheoreticalEmpiricalModel* occurrence of the *PM_Template*	Instance
hasSM	Determines the *SurrogateModel* occurrence of the *PM_Template*	Instance
is_Solved	Determines the *cDSP_Template* occurrence of the *DSE_Template*	Instance
hasPSA	Determines the *PSA_Template* occurrence of the *DSE_Template*	Instance
hasPM	Determines the *PM_Template* occurrence of the *DSE_Template*	Instance

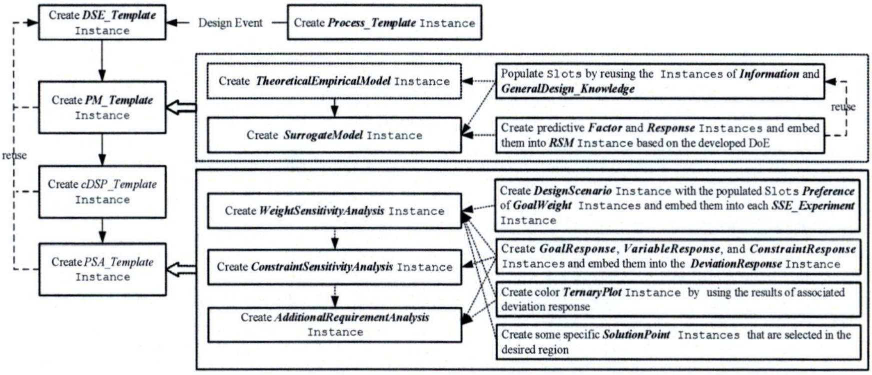

Fig. 6.6 DSE process template instantiation process [29]

(3) Exploration Instantiation Employing DSE Process Template Ontology

According to the procedure for DSE defined in Sect. 6.2.2, DSE process template comprises the PSA, cDSP, and PM templates, as illustrated in Fig. 6.6. Before initializing the DSE process template, the designer needs to determine the corresponding design event defined in the PEI-X diagram process, and regulate the examined design problem's relevant knowledge and design information. In this section, we focus on creating and populating the PM and the PSA template, with the corresponding instantiation process listed below.

- **Create *PM_Template* `Instance`**. Exploit the input instances of the `Classes` *Information* and *GeneralDesign_ Knowledge* determined in the design event to create and populate the *TheoreticalEmpiricalModel* `Instance`. If some existing TEM instances are unavailable, a *SurrogateModel* `Instance` is created, i.e., create the predictive *Factor* and *Response* `Instances` and embed them into the *RSM* (Response Surface Model) `Instance` based on the developed DoE. The newly created template instance of the surrogate model is stored as new knowledge for subsequent reuse.
- **Create *PSA_Template* `Instance`**. The PSA template involves weight and constraint sensitivities, and additional requirement analysis models that depend on the problem examined and are populated into the PSA template instance `Slots`. The WSA `Instance` is an essential module supporting the designer to determine the anticipated solution region. The WSA input `Slot` module is the experiment of the solution space exploration (*SSE_Experiment*), and the output `Slots` are the *TernaryPlot* and *DeviationResponse* sub-modules, affording insight for the designer during the decision-making process. The DesignScenario Instance is created based on the populated slots Preference of the GoalWeight Instances and is embedded into each SSE_Experiment Instance at the initial stages of the post-solution analysis. The cDSP template results are captured by the instances of

the Classes *GoalResponse*, *VariableResponse*, and *ConstraintResponse*. Meanwhile, these various types of response instances are populated into the *DeviationResponse* Instance. The Instance of color *TernaryPlot* per system goal is created exploiting the associated goal deviation response results, which is populated into the WSA module, and then based on these ternary plots, a common region satisfying all the system goals is established, forming the superimposed ternary plot.

In some special problem cases, e.g., if the initial design solution space has no common region, the designer must perform a detailed post-solution analysis to increase understanding of the design response and the confidence in prediction. Therefore, the CSA and ARA module Instances are established to capture the information reusability in the design space adjustment for a satisfying range. In the CSA module, the extra capacity of the design space is identified to adjust the active constraints. While, in the ARA module, the variables/constraints are further analyzed as an extra system goals requirement. Additionally, the Instance of *TernaryPlot* per variable/constraint is created by employing the associated variable/constraint deviation response results. All information from the ARA, CSA, and WSA modules embedded into the PSA template instance contribute to determining the desired solutions from which some specific *SolutionPoint* Instances are chosen to support the designer in identifying the design set points. Based on the scenarios defined in Sect. 6.2.3, the different instance versions can document the modified information based on the deviation response. For example, considering the target requirements to be accepted or approved, the acceptable value is modified based on the designer's experience, knowledge, or preference to obtain a satisficing common region. The adjusted acceptable value is captured by the various *TernaryPlot* Instance versions embedded into the corresponding ARA, CSA, and WSA modules.

6.2.5 Test Example: Designing of Hot Rod Rolling Process Chain

In this section, we illustrate the DSE process template ontology utility by considering the design problem of an automotive gear manufacturing process: a complex system design requiring several decisions to be made. Manufacturing automotive gears involves several processing stages. This section primarily focuses on the hot rod rolling (HRR) process stage.

(1) **Designing the Hot Rod Rolling Process Chain**

The design problem of an HRR process chain is presented in Chap. 4. In this section, we frame a boundary within the problem as determined by Nellippallil et al. [28], and illustrate the information reusability during the design space exploration procedures employing the DSE process template. We demonstrate how a designer can capture, represent, and document reusable information of the hot rod rolling problem and thereby support designers making decisions considering robustness in their design.

(2) Populating a Basic DSE Process Template Instance

Based on the DSE procedure of Sect. 6.2.2 and the DSE process template instantiation approach, we create a primary DSE process template instance and illustrate the populated sub-templates for the problem model and the corresponding post-solution analysis by exploiting the cooling module of the HRR problem.

- **Creating and Populating the Process Template for the Problem Model**

The problem model templates assist the designer in determining the initial design space and providing a combination of the design information as inputs for the cDSP model. In other words, the process designer initially determines the basic design space features during the creation of the exploration processes (showcased in Fig. 6.7). For the HRR process chain problem, the reader is referred to the embedded `Instance` "ProblemModel-1" illustrated in window "①" of Fig. 6.7. The problem module inputs involve chemical compositions, e.g., manganese concentration after rolling [Mn], carbon concentration [C], final austenite grain size after rolling (D), cooling conditions, i.e., cooling rate (CR). The outputs are the end product's mechanical properties, i.e., hardness (HV), tensile strength (TS), and yield strength (YS) for the rod, which dependend on the final microstructure after cooling, e.g., the pearlite interlamellar spacing (S_0), the phase fractions of ferrite (X_f) and pearlite ($1 - X_f$), and

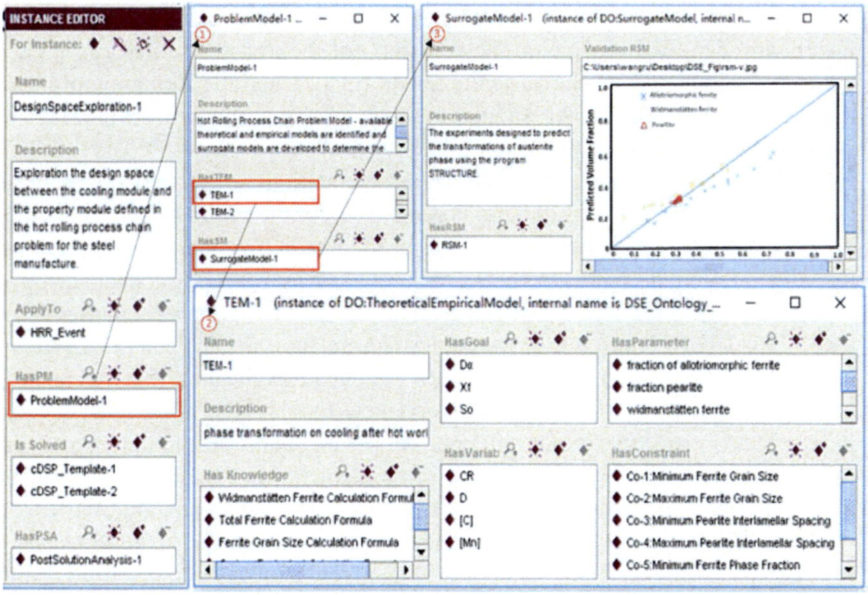

Fig. 6.7 PM template instances embedded in the DSE process template [29]

the ferrite grain size after cooling (FGS, D_α), along with the composition variables like manganese ([Mn]), phosphorous (P), nitrogen ([N]), and silicon ([Si]).

Within the boundary of the problem, the cooling and property modules in HRR are addressed through two compromise Decision Support Problem (cDSP) mathematical constructs relying on the design set points of the steel manufacturing process. Thus, the designer creates two theoretical and empirical model (TEM) modules providing a combination of design information as inputs for the cDSP models, i.e., "TEM-1" and "TEM-2". As presented in Fig. 6.7, the design information comprising the module includes: "system goal," "constraint," "system variable," "design parameter," and "existing knowledge" about the available functional relationships. For further details on these instances, the reader is referred to [28]. For example, the "TSM-1" embedded in the `Instance` "ProblemModel-1" is shown in window "②" of Fig. 6.7.

The HRR problem examined here requires predicting the transformations of the austenite phase. Depending on the cooling criteria, the transformation phase during cooling converts austenite to various steel phases, i.e., allotriomorphic ferrite, pearlite, widmanstätten ferrite, bainite, and martensite, etc. [39]. It is essential to predict these transformation phases to manage the banding phenomena occurring in the microstructure. To develop surrogate models for the different steel transformation phases, a meta-modeling approach is employed (window "③" in Fig. 6.7). In this case, it is assumed that the austenite transformations only occur in the ferrite and pearlite phases.

Window "①" of Fig. 6.8 presents a three-level fractional factorial design to develop the response surface models for the austenite transformation to pearlite and ferrite through the embedded `Instance` of "RSM-1". For the experimental design, four factors are identified to develop the responses of the transformation stages due to the massive influence on the austenite transformation and the formation of banded microstructures [40]. For the simulations, we also identify the factor values that correspond to the appropriate factor levels (window "②" of Fig. 6.8). The simulation runs on programs that attain the input–output correlations so that the cDSP is formulated based on the specific problem. For instance, for the problem of [36], the finite element simulation software ABAQUS is used, in which a finite element model for the hot rod rolling is created to predict the oval to round geometry conversion during rolling. In this experiment, the experimental runs predicting the steel phases exploit the "STRUCTURE" program based on the data and tools available in [39]. The input and output datasets are utilized to estimate the parameter values of the meta-model based on least squares. Usually, a regression meta-model is one of the three types: (1) main effects model (first-order polynomial), (2) main and interaction effects (first-order polynomial augmented with two-factor interactions), (3) quadratic model with quantitative factors (second-order polynomial including purely quadratic). In Fig. 6.8 window "③", the regression model created for fraction pearlite X_p with the lowest sum of squares, an error R^2 value of 0.99 is given, and it is generated by fitting a second-order polynomial function.

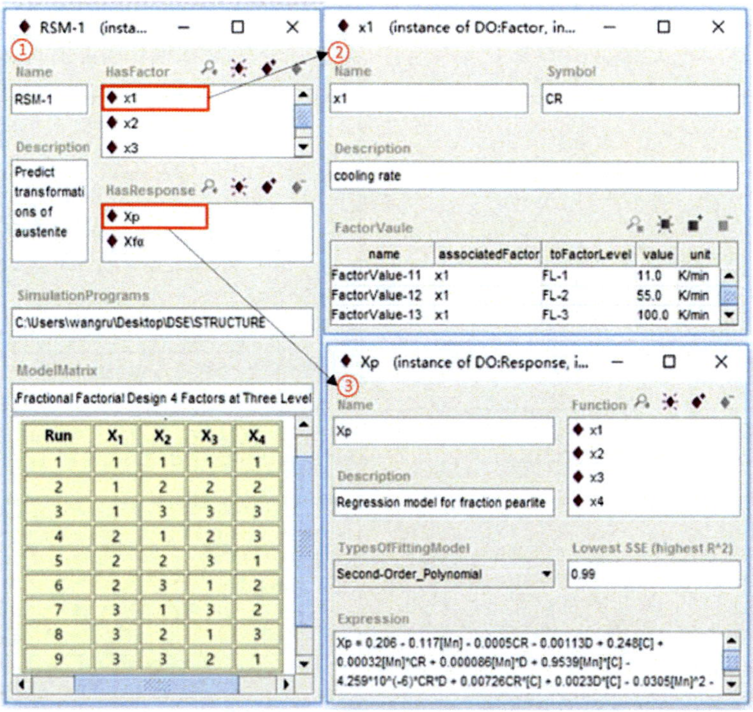

Fig. 6.8 Instance of the RSM model [29]

- **Creating and Populating Templates for Post-Solution Analysis**

According to the design information combination generated from the specific problem model, we formulate two cDSP templates that are utilized to determine the design variable values, which satisfy a set of conflicting goals, i.e., maximizing *YS* and *TS* for the end mechanical properties of the rod and minimizing D_α and S_0 for the microstructure space after cooling. For further details on the cDSP formulations, the reader is referred to [28], while the description on creating and populating the cDSP template is presented in [4]. In this section, we focus on preserving flexibility in identifying design solutions under uncertainty and on the process where a designer must explore several design preferences to guarantee a "satisficing range" solution and trade-off the multiple conflicting objectives. The sensitivity analysis and deviation response information during the exploration process is extracted from the `Slots` of the *PSA_Template.*

In Fig. 6.9, we present the weight sensitivity analysis to determine the desired solutions satisfying high priority goals. In this figure, the deviation function is an *Archimedean* formulation affording the designer to explore as many scenarios as possible by assigning several weight combinations to the associated system goals. For this scenario, the process designer establishes four exploration experiment types, which are captured by the `Slots` "Input" in the "WeightSensitivityAnalysis-1"

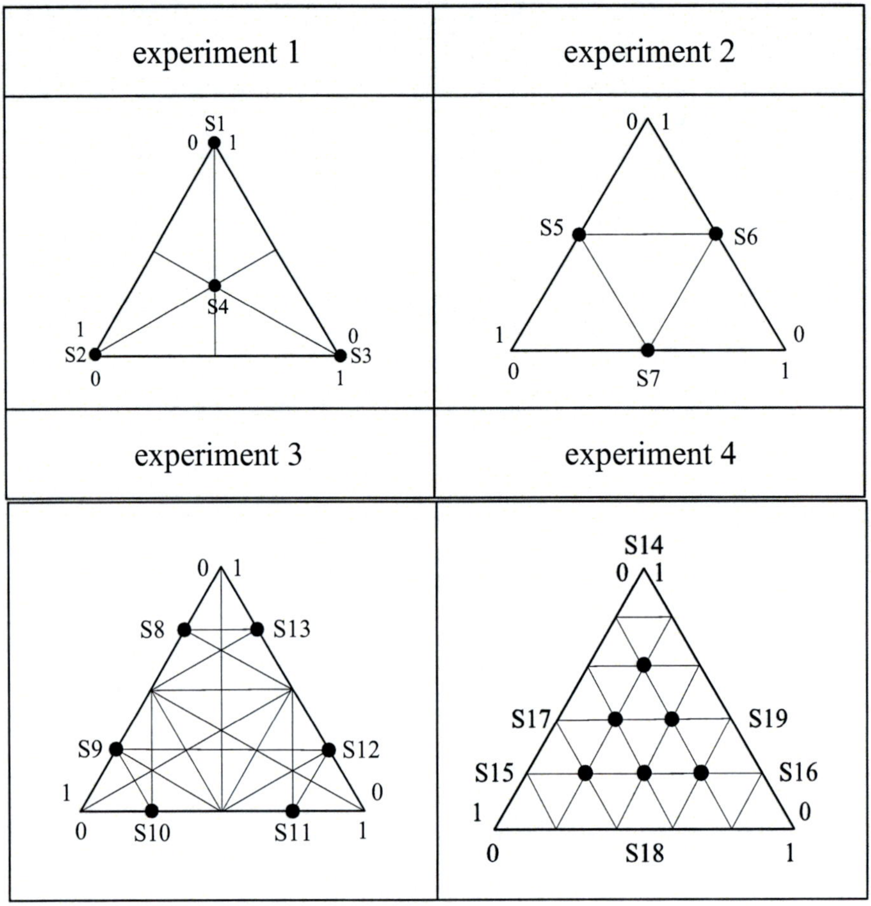

Fig. 6.9 Scenarios for the solution space exploration trials [29]

(window "①" in Fig. 6.8) of the "cDSP_Template-1". The latter is used to define the microstructure factors after rolling and the operating set points for cooling that satisfy the system goals D_{α}, X_f, and S_0 as defined by the system variables CR, D, [C], and [Mn].

As illustrated in Fig. 6.9, the design scenarios 1–4 in "Experiment-1" consider the case where the designer's interest is either achieving the target of one of the system goals (S1, S2, and S3) or giving equal preference to all the goals considered (S4). The design scenarios 5–7 in "Experiment-2" consider two goals given equal preference, while the third goal is not given any preference. The design scenarios 8–13 in "Experiment-3" refer to the case where the designer has a greater preference for one goal, less preference for the second goal, and no preference for the third goal. Design scenarios 14–19 in "Experiment-4" consider all the goals are given preferences with two of them having the same preference. The design scenario's preference value per goal weight is presented in window "②" of Fig. 6.10. The cDSP template is

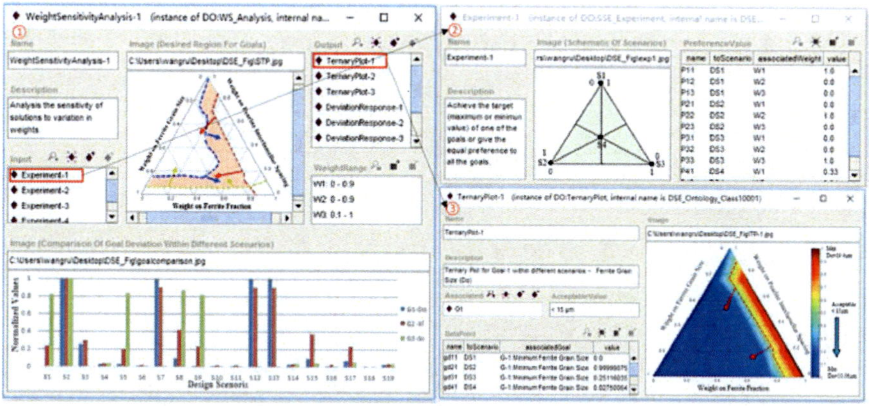

Fig. 6.10 Instance of weight sensitivity analysis for cooling module [29]

used on various design scenarios exploiting the DSIDES tool. The cDSP models are executed to minimize the deviation function and produce the corresponding system variable values. Then, window "③" of Fig. 6.10 presents the system's goal deviation variables representing the degree that a value is away from the target. Based on those deviation variables, ternary plots are used to visualize and explore the solution space, e.g., the solution space for "G1" (minimizing the ferrite grain size D_α) is illustrated in Fig. 6.10. From the latter figure, the designer identifies the minimum D_α value by employing the current cDSP template configuration information of 10.06 μm satisfying the acceptable value from the existing empirical knowledge of 15 μm. The contour region determined by the red dashed lines comply with the "G1" design requirements. The similar ternary plots for all system goals are generated in the `Slots` "Output" of `Instance` "WeightSensitivityAnalysis-1". According to those design datasets, all goals are superimposed in a single plot to determine a common region (pink area in window "①" of Fig. 6.10), meeting all goals and adding confidence to the designer's decision-making process. Meanwhile, the weight ranges linked with the joint region are also determined, and any weight combination that sums to one guarantees the desired solution. A bar chart compares the goal deviation within different design scenarios to enhance a designer's understanding of the solution space. In this bar chart, a shorter bar indicates that a solution deviation from the target is low. Thus, a better design point/solution is found. Observing and analyzing the pink area (superimposed region), we predict five solution points in the following design scenarios: S6, S10, S11, S16, and S18. Thus, the designer must consider the design trade-offs based on specific requirements and choose the final design among the five solution points.

To enhance the reader's understanding of this process, seven points are selected to compare good and bad solutions considering different scenarios from the common region identified, its boundary, and outside it (Fig. 6.11). The design point information is populated in the `Slots` "Results_SSE" (results of solution space exploration)

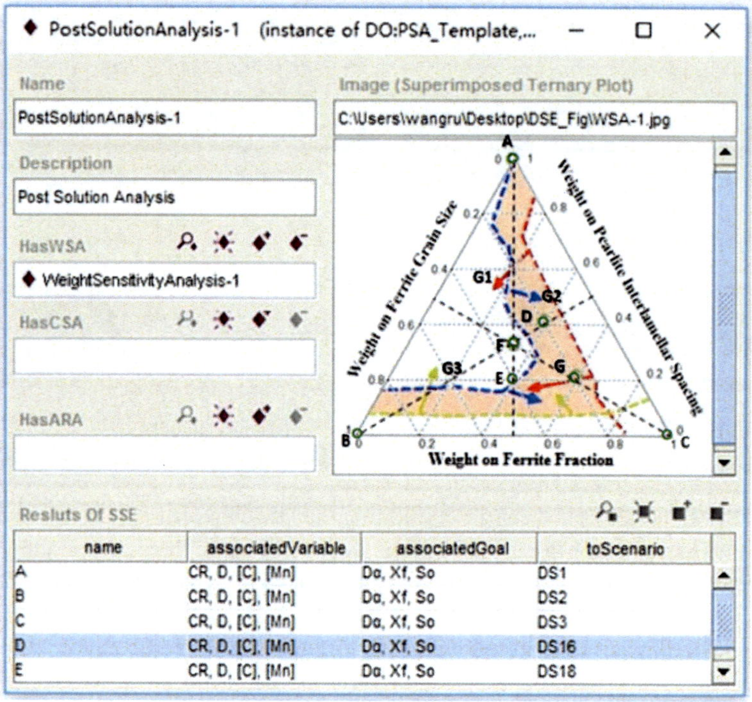

Fig. 6.11 PSA template for cooling module instance [29]

of `Instance` "PostSolutionAnalysis-1". The detailed results are presented in Table 4.8.

For Table 6.5, solution points A, B, and C satisfy the associated goals, i.e., minimum ferrite grain size (D_α), maximum ferrite fractions (X_f), and minimum pearlite interlamellar spacing (S_0), respectively. Compared to the other design points E, F, and G, point D that lies in the joint region and correspond to the design scenarios S16 and satisfy all the conflicting goals in the best possible manner. Thus, point D

Table 6.5 Results on the selected points [29]

Sol Pt	D_α	X_f	S_0 μm	*CR* K/min	*D* μm	[C] %	[Mn] %
A	12.5	0.684	0.149	11	30	0.19	1.02
B	10.06	0.681	0.176	99.9	30	0.18	0.7
C	19.9	0.714	0.182	11	74.2	0.18	0.7
D	10.74	0.681	0.151	44.4	30	0.18	0.94
E	10.33	0.673	0.151	70.3	30	0.18	0.93
F	10.33	0.673	0.151	70.1	30	0.18	0.93
G	11.05	0.687	0.151	33.06	30	0.18	0.95

is the recommended solution. Then, this information passes to the following cDSP models formulated for subsequent manufacturing operations, affording a horizontal manufacturing process chain integration.

(3) Populating a Special DSE Process Template Instance

We create a primary DSE process template instance by employing the PM and the PSA templates to populate the design space exploration reusability information for the cooling module in HRR. For that scenario, a common region exists, satisfying all goals simultaneously during the post-solution analysis. Thus, the process designer confidently identifies the design set points from the desired solution space to meet the target microstructure requirements. However, the case where a common region does not exist is also possible. This scenario is discussed in this section by instantiating a particular DSE process template.

In the HRR problem, the module after the microstructure correlation calculation (cooling module) predicts the mechanical properties, with the corresponding system goals for the rod (end product) are hardness (HV), tensile strength (TS), and yield strength (YS). The theoretical and empirical models of the property module (TEM-2 `Instance`) are populated in the PM template instance, allowing the designers/users to define the design elements, e.g., variables, constraints, and goals, along with the mathematical models considered in the cDSP model (cDSP_Template-2 `Instance`) used to solve the property module. Similar to the exploration processes presented in the previous section, the basic PSA template module "WeightSensitivityAnalysis-2" `Instance` is established, and its output `Slots` are created according to the cDSP_Template-2 `Instance` results which are obtained by performing the solution space exploration scenarios. Based on the ternary plots of each system goal, i.e., the rod's mechanical properties, created by utilizing the associated goal deviation response results, a superimposed ternary plot is created to assist the designer in determining the desired solution region, which satisfies the requirements (Fig. 6.12).

In Fig. 6.12, the contour region identified by the blue dashed lines satisfies the system goal-1 of maximizing yield strength, and the maximum yield strength achieved is 320.6 MPa when the weight assigned to yield strength goal is 1.0. The pink contour region determined by the orange and green dashed lines simultaneously satisfies the system goals to maximize the hardness and tensile strength, which are achieved when the associated goals weight tend to one. The maximum tensile strength value is 750 MPa, and the hardness is 170. However, if the designer adjusts the acceptable target values, a common region satisfying all system goals does not exist. In this case, to make a design decision, the process designer must consider additional requirements to adjust the initial design space and exploit the associated information. When included in the solution space exploration scheme and the system goals, the information linked with the system variables and constraints will/could provide the designer with information that can be exploited to make a design decision in such cases. In the following section, the same case is explained for the HRR problem.

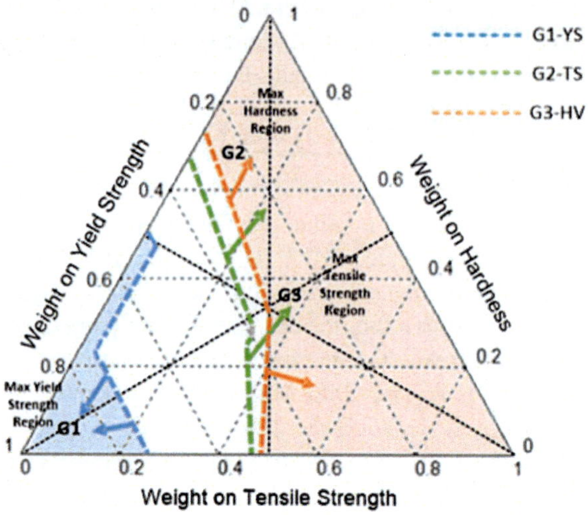

Fig. 6.12 Superimposed ternary plot [29]

In the HRR problem, there are other design requirements affecting the mechanical properties, i.e., the banded microstructure after cooling and the material's impact toughness. To measure the impact on toughness, the transition temperature (ITT) is used to represent the boundary between brittle and ductile failure subjected to impact loads, and pose a constraint on the initial design space. Meanwhile, we study the banded microstructure management after cooling by considering the pearlite fraction (S_0) and ferrite fraction (X_f) after cooling, which are defined as a system goal in the previous process stage (i.e., cooling module) and system variables in the "TEM-2" instance, respectively. In the mechanical properties module that considers post-solution analysis, the `Slot` of additional requirement analysis must be populated after the "WeightSensitivityAnalysis-2". The "AdditionalRequirementAnalysis-2" `Instance` is established based on the deviations of the constraint (impact transition temperature) and the system variable (ferrite fraction) of the problem examined. Window "②" in Figs. 6.10 presents the solution space ternary plots considering the constraint in ITT and the system variable (ferrite fraction (X_f)) with respect to the weight change linked to the system goals defined by hardness (HV), tensile strength (TS), and yield strength (YS).

The red dashed lines identify where the impact transition temperature is minimum in the constraint solution space, while the red dashed line refers to zero *ITT*. In the variable solution space, white and gray dashed lines define the contour regions of higher pearlite and ferrite fractions, with the intermediate region defining the highly banded microstructure that has both pearlite and ferrite. By comparing plots, we find that the value increases (65–100 °C) as the pearlite fraction increases, which is unacceptable in practical designs. Therefore, we add a green dashed line referring to *ITT* of 65 °C. Here, we aim to achieve a maximum value of ferrite fraction and a minimum value of ferrite grain size and ultimately afford microstructure banding. These additional system goals and requirements are presented in the superimposed ternary plot (window "①" of Fig. 6.10) to assist designers during the trade-off choices while aiming for a satisficing decision. The pink contour region having a high ferrite fraction is determined in a compromised manner. The pink contour region meets the impact transition temperature and yield strength requirements and compromises the hardness and tensile strength requirements. To further illustrate this process, some particular design points are selected (window "①" of Fig. 6.10). Finally, the recommended solution of interest is point B, as it has the maximum yield strength and the highest ferrite fraction.

6.3 Ontology-Based Uncertainty Management in Designing Robust Decision Workflows

6.3.1 Requirements for Uncertainty Management in Decision Workflows

Leveraging knowledge related to the design process is essential to improve an enterprise's agility, and thus strategic methods to effectively manage an enterprise's intellectual capital have drawn significant attention [41]. The design processes of complex engineered systems are in their majority *ad-hoc* structure, exploiting previous design experience [2]. Hence, the quality of a system design process greatly influences design and cost efficiency [29]. According to Simon [42], "design process strategies can affect not only the efficiency with which resources for designing are used but also the nature of the final design as well". Hence, designing the design processes, i.e., meta-design, is crucial in a systems-based design strategy employed to design complex systems, especially for multiscale systems with high degrees of uncertainty and nonlinearity [43].

Various uncertainties in design coexist, and from a model-based complex engineered systems realization perspective, a computational environment enabling the system model to integrate the vital design information is necessary to manage various uncertainties by applying existing processes, methods, and tools [8]. Thus, given the semantic ontology interoperability in the computer environment, in Sect. 6.4, we present a template-based ontological method to characterize and encapsulate the uncertain knowledge in the engineering design on the foundation of Decision Support Problem Technique (DSPT) introduced below. The developed ontology is aimed for a robust decision process enabling human designers to understand and predict the process activities during the decision-making.

Considering DSPT, a robust decision-making process involves a specific set of methods that shall help human designers identify potential robust strategies under the conditions of complexity and uncertainty. A decision-centric meta-design of a complex system design needs the decision process information flows to be efficiently combined and organized, affording a human designer to accommodate uncertainty and make a robust design decision. Standard design process models, e.g., IDEF0, BPMN, and EPC are not adequate to describe the information related to uncertainty existing in a process chain. Inspired by this, Phadke [44] develops a P-Diagram based on a semantic graphical representation scheme representing the process/product's quality features that are valuable to describe a robust design task. From the DBD perspective, the PEI-X diagram can visualize hierarchical decision processes, providing a basis to graphically represent the robust design information amid decision-making actions. Nevertheless, it is not trivial to utilize the PEI-X and P-Diagram to express the activities linked to robust design and the uncertainty impact on the decision-making processes. Therefore, it is necessary to use a hierarchical process model with a stronger semantically graphical expression to explicitly depict the value of the parameters interlinked with individual subsystems and the propagation characteristics of the uncertainty in the model and the process chain.

6.3.2 *Procedure for Designing Robust Decision Workflows*

In this section, we propose a domain-independent method to help designers define and create computational and reusable decision workflows, which involve experimental designs, statistical analysis, and decision-making. The suggested method has the following stages:

STEP 1: Identifying the uncertainty types

It is clarified and verified in several research studies that early design phase variations significantly impact the performance and quality of the subsequent design. Hence, while effectively managing the design uncertainties, the most crucial part is identifying their types that require determining whether the uncertainty can be quantified. Thus, while specifying the design problem, the primary task is to distinguish the design parameters between numeric and attribute types. Four attribute types exist defining the simulation model parameters: control factors, noise factors, response, and fixed parameters. The numeric parameters must be defined as intervals or discrete values, as the system uncertainty involves the attribute types. To afford the designers completing this work, we define a graphical expression to design the robust design hierarchies, utilizing strong semantics representing each element's features in the robust design model layer. In Table 6.6, we present the revised graphical expression of [45] that captures three semantic information items of the model entity, data attribute, and composite pattern. This graphical representation facilitates identifying the uncertainty management types and represents the robust design hierarchy, increasing the understanding of the relationships among simulation model elements and uncertainty propagation.

To quantify the uncertainty sources, we define three variations based on a system's parameter distribution probability: namely, mean (μ), deviation (Δx), and variance (σ^2). Acquiring these quantifiable variation values will determine the system's response estimation that will be managed using two methods:

Table 6.6 Graphical expression for the hierarchies in robust decision workflows

Entity	Symbols	Semantics
hexagon box		uncertain model
pentagon box		certain model
rhombus box		control factors, noise factors, the response, and the fixed-parameter
Arrow Line	**Symbols**	**Semantics**
double dotted line		ranged & uncertain
double solid line		ranged & certain
single dotted line		discrete & uncertain
single solid line		discrete & certain
Composite Pattern	**Symbols**	**Semantics**
arrow out of the box to the bottom	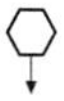	output response
arrow out of the box to the right	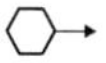	determined variable
arrow from the top into the box		goal/required response
arrow from the left into the box	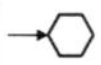	given value/parameter

- Input Parameter Uncertainty (IPU). The system model handles the uncertainty imposed by the input parameters variability, i.e., control and noise factors. This method guarantees that the solutions are insensitive to the input parameter variation. The corresponding robust design for this type of uncertainty management type is Type I (for noise factors) or Type II (for control factors).
- Model Parameter Uncertainty (MPU): Handling the uncertainty imposed by the unparameterizable variability in the system model. The corresponding robust design is Type III, and it is used to identify a ranged set of solutions that are relatively insensitive to the variability within the system model.

Due to the limited system knowledge, some simplifications and assumptions of the simulation require the uncertainty propagate in a chain of models. This uncertainty type is unquantifiable and is managed by employing the Model Structure Uncertainty (MSU) method. MSU belongs to Type IV robust design and includes two other uncertainty management types. Several types of robust design will be performed in Step 3.

STEP 2: Designing the hierarchical decision workflows

The robust design problem is structurally modeled using the graphical expression defined in Step 1, with the majority of the information describing the specific problem being organized in a displayable manner. A further issue that needs to be solved is exploiting this information to solve the design problem through a reasonable process. Considering DSPT, various computational task sequences related to decision-making are organized within the decision workflows. The process granularity of the different levels is partitioned and planned to form the decision workflows meta-design hierarchy. Based on the identified uncertainty management methods and the robust design types, we define the hierarchical decision workflows to provide executable processes for obtaining solutions satisfying the design requirement in the given design space. To increase the effectiveness and efficiency of the meta-design of decision workflows, we model a graphical decision-making procedure based on the PEI-X process language format. Subject to the functional goals, the decision workflows comprise five process templates types: "**Design**", "**Phase**", "**Event**", "**Decision**", and "**Task**" [41]. The iteration of the specific activities is applied in each executable process template through the embedded computable module. The generated information is exploited to manage the uncertainty in Step 1.

STEP 3: Executing the sequential computability routines

Step 2 is to partition and plan the executable design activities. To obtain a solution that meets the design requirements, the sequential computable routines embedded in the process template must be executed. Defining the computable routines in the decision workflows is related to the design goals. Ref. [29] describes some of the goals related to the design space exploration, e.g., the design preferences exploring in the Post-Solution Analysis (PSA) module and the response surface modeling in the Design of Experiments module. In this section, we focus on other computable routines associated with the three types of uncertainty management identified in Step 1, namely:

- *The computable routines for the IPU management.* The response's mean and variance formulations defined in [46] are utilized in the robust designs Type I and Type II for the IPU management. Solutions satisfying a set of performance requirement targets are determined by considering the system's performance deviation resulting from the variation in control and noise factors. The Design Capability Indices (DCI) must be calculated in the robust design Type III because the design variables have adjustable ranges rather than a single value. Hence, the solutions satisfy a ranged set of performance requirements [47].
- *The computable routines for the MPU management.* In the robust design Type III, the premise of the above design scenario is that the simulation model is certain. However, if the simulation model is identified in Step 1 as uncertain, the Error Margin Index (EMI) must be calculated relying on the variability interval formulation in the system response. Choi et al. [48] suggest a method estimating the lower/upper response.

- *The computable routines for the MSU management.* All the previous computable routines may appear in the robust design Type IV due to the uncertainty type identification for each model in the simulation models chain. A designer must evaluate discrete points using the simulation model to identify the adjustable ranges for the given design space while considering uncertainty propagation. The most crucial phase of the evolution process is searching the feasible region for the interdependent space between two models based on the hyper-dimensional EMI (HD-EMI) and employing the discrete points created from the maximum, minimum, and mean response functions. For a detailed calculation developed in the IDEM, the reader is referred to [49].

6.3.3 Ontology for Designing Robust Design Decision Workflows

According to the uncertainty management requirements of Sect. 6.3.1, we define a modular template to achieve the goals of reusability and executability during the robust design decision process. Furthermore, we develop a frame-based ontology relying on the module elements embedded in the robust design template that enhance the designer's understanding of the process behavior. Then the ontology-based template instantiation procedure is elaborated to maintain the robust decision workflows' design approach. However, instead of employing visual tools to display the ontology structure, such as OWL-VisMod, the meta-design visualization in robust decision workflows exploits a graph-based editing tool in Protégé [50].

(1) **The Modular Template for the Robust Design**

In a computational environment, a modular-based design approach affords a designer the opportunity to construct executable and reconfigurable process templates that can implement flexible configurations for the different uncertainty management types identified by the suggested method presented in Sect. 6.3.2. Hence, we develop a modular-based template to enhance the executability and reusability capabilities that are useful for a robust design process.

In Fig. 6.13, we illustrate that the robust design template is visualized as a structure similar to a printed board assembly with electronic components, in which the elements are represented by "chips," and the procedure to achieve the robust decision workflows design is the "breadboard". The three reuse scenarios of [41] and several templates and modules from [29, 41] are reused to capture the decision process information and the related problem model. The ***Process Template*** that integrates the decision workflows meta-design is reused in the form of an assembly, and the ***Process Template*** involving specific information of the "chips", e.g., the ***Support Problems*** or the ***cDSP template***, is utilized to populate the related property `Slots` of the robust template. For example, the parameter information of the control or noise factors, the related responses, and fixed parameters of the simulation mode are captured and documented by the ***Response Surface Model*** module developed in

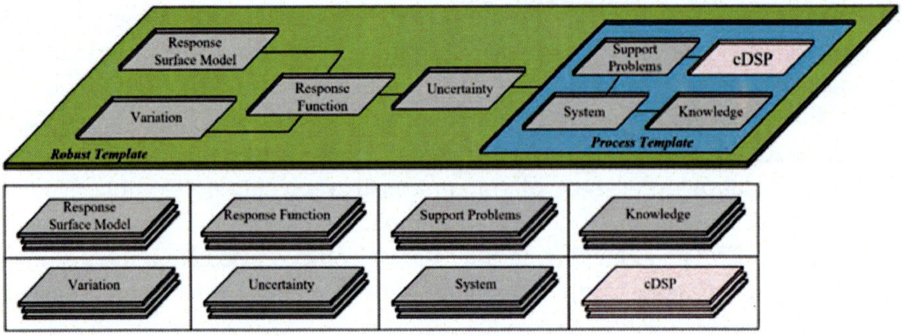

Fig. 6.13 The modular template for the robust design [51]

the design space exploration ontology. In this section, we define other new modules related to uncertainty management: ***Response Function***, ***Variation***, and ***Uncertainty***, with the corresponding functions presented in detail in Sect. 6.3.2.

(2) **Defining Classes and Slots**

We develop a frame-based ontology that is appropriate for the robust design decision process template using Protégé 3.5. The ontology can capture and document the reusability information in the robust design and support the integrated uncertainty management in the design process. The robust design decision ontology comprises the `Classes` and `Slots`, and by mapping to the robust design template, the "chips" in the "breadboard" determine the main ontology `Classes`. Meanwhile, some `sub-classes` are also identified, increasing the semantic richness and integrity of the robust design template ontology. Here, the focus is on defining `Classes`: *Uncertainty*, *Variation*, and *ResponseFunction*, and the semantic relationships captured using `Slots` among those `Classes`. Two `Slots` types exist: data and object slots. The former link the classes to the end data, while the latter links classes to other classes. Detailed definitions of `Classes` and `Slots` are presented in Tables 6.7, 6.8 and 6.9. It should be noted that some `Classes` and `Slots` that are reused from previously developed ontologies [4, 29, 41], such as *Process_Template*, *hasParameter*, *name*, *value*, etc., are not described again.

(3) **Instantiation Procedure of the Robust Decision Process Template**

Based on the design method for a robust decision workflow introduced in Sect. 6.3.2, the robust design template comprises the following modules and templates: ***Process Template***, ***Uncertainty***, ***Variation***, and ***Response Surface Model***, as presented in Fig. 6.14. In the robust design template instantiation process, a crucial aspect is creating a PEI-X process template managing a proper granularity solving the defined design problem under uncertainty. The method to create and populate the property slots in the process template is described in [41]. This work highlights the robust decision workflows design and achieves uncertainty management by reusing the `Instances` *Information* created in the *Process_Template*. In the suggested

Table 6.7 Classes of the robust design process ontology [51]

Class	Definition
Robust_Template	A formulation integrating all the associated modules and representing the robust design information structure
Uncertainty	A class representing the different management types under the various source of the uncertainty types
ResponseFunction	A class representing the system's performance functional relationship response under uncertainty
Variation	A class representing the quantitative degree of the factors' variation in the uncertainty
InputParameterUncertainty	A sub-class of "*Uncertainty*" representing the robust design information management due to uncertainty from the system input parameters (i.e., "control factors" and "noise factors")
ModelParameterUncertainty	A sub-class of "*Uncertainty*" representing the robust design information management due to uncertainty from the system simulation model parameters (i.e., the unparameterizable variability of the model)
ModelStructureUncertainty	A sub-class of "*Uncertainty*" representing the robust design information management due to uncertainty from the system model structure formulation (i.e., the approximations and simplifications in a model)
MeanResponseFunction	A sub-class of "*ResponseFunction*" representing the response's mean function under different uncertainty types
VarianceResponseFunction	A sub-class of "*ResponseFunction*" representing the response's variance function
LowerResponseBoundFunction	A sub-class of "*ResponseFunction*" representing the response's lower deviation function
UpperResponseBoundFunction	A sub-class of "*ResponseFunction*" representing the response's upper deviation function
Mean	A sub-class of "*Variation*" representing the factors' given mean value (including noise and control factors)
Variance	A sub-class of "*Variation*" representing the factors' given variance value (including noise and control factors)
Deviation	A sub-class of "*Variation*" representing the control factors' given variance value

ontology, the instantiation procedure for the robust decision workflows involves three major phases:

- Creating the *Robust_Template* `Instance` and the related modules, i.e., *Uncertainty, Variation,* and *ResponseFunction.* Based on the parameter features of the defined design problem, i.e., the parameters variation in the simulation model, the designer chooses and edits the relevant composite pattern, arrow lines, and boxes to illustrate the graphical hierarchies of the simulation model by employing the

Table 6.8. Object slots of the robust design process ontology [51]

Class	Definition	Type
robustDesignType	Specifies the robust design type	Instance
hasSM	Specifies the *Robust_Template* surrogate model instance	Instance
hasTEM	Specifies the *Robust_Template* theoretical, empirical model instance	Instance
hasCFs	Specifies *Robust_Template* the control factor instance	Instance
hasNFs	Specifies the *Robust_Template* noise factor instance	Instance
hasUncertainty	Specifies the *Robust_Template* uncertainty instance type	Instance
hasMeanResponseFunction	Specifies the *MeanResponseFunction* instance of the *InputParameterUncertainty* and the *ModelParameterUncertainty*	Instance
hasVarianceResponseFunction	Specifies the *VarianceResponseFunction* instance of the *InputParameterUncertainty* and the *ModelParameterUncertainty*	Instance
hasUpperResponseBoundFunction	Specifies the *UpperResponseBoundFunction* instance of the *InputParameterUncertainty* and the *ModelParameterUncertainty*	Instance
hasLowerResponseBoundFunction	Specifies the *LowerResponseBoundFunction* instance of the *InputParameterUncertainty* and the *ModelParameterUncertainty*	Instance
MeanOfCF	Specifies the control factor's *Mean* instance	Instance
MeanOfNF	Specifies the noise factor *Mean* instance	Instance
VarianceOfCF	Specifies the control factor *Variance* instance	Instance
VarianceOfNF	Specifies the noise factor *Variance* instance	Instance
deviationOfCF	Specifies the control factor *Deviation* instance	Instance
designSpace	Specifies the generated discrete points for the defined design variables instance in the design space	Instance
interdependentSpace	Specifies the generated discrete points for the defined design variables instance in the interdependent space	Instance

graphical expression defined in Step 1. This identifies the robust design types and the involved uncertainty management.

- Creating the *Process_Template* `Instance` and the related modules, i.e., *Interface, GeneralDesign_Knowledge, Information, cDSP_Template, Sys_Entity,* and *SPs_Entity*. According to the identified robust design type, the designer selects and edits the relevant interface, information, and process entity to illustrate the

Table 6.9 Data slots of the robust design process ontology [51]

Class	Definition	Type
lowerRequirementLimit	The lower requirement limit value for the system performance design requirement	Float
upperRequirementLimit	The upper requirement limit value for the system performance design requirement	Float
targetForDCI	The design capability index target value	Float
targetForEMI	The error margin index target value	Float
targetForHD-EMI	The hyper-dimensional error margin index target value to estimate interdependent space	Float
valueOfHD-EMI	The hyper-dimensional error margin index value to estimate design space	Float
CL(1-α)	The confidence level value to predict the interval estimation with the Student t-distribution	Float
numOfPredictors	The predictors' cardinality in an approximate model to predict the interval estimation with the Student t-distribution	Integer
numOfSamples	The samples cardinality to predict the interval estimation with the Student t-distribution	Integer
decisionCriterion	The decision preference to indicate the location of the system response	String

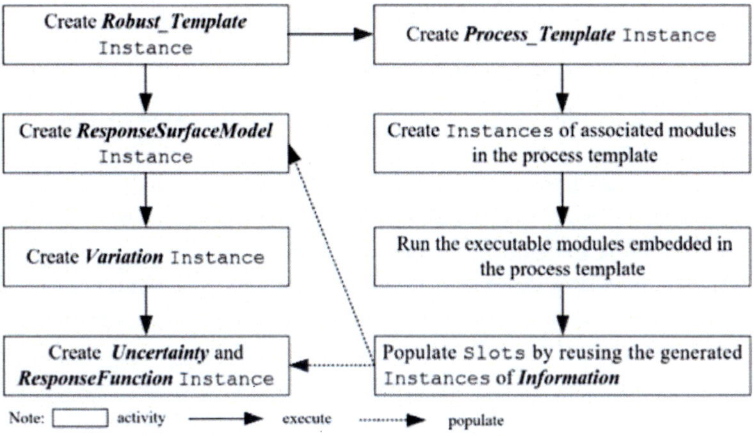

Fig. 6.14 Instantiation procedure of the robust design template [51]

graphical hierarchical decision workflows that primarily involves partitioning and planning the support problems.

Running the executable modules embedded in the defined decision workflows and populating the property `Slots` in the *Robust_Template* `Instance` by reusing the created *Information* `Instance`. In the *Process_Template* `Instance`, the executable computing modules, e.g., solving the cDSP-DCI/EMI model, calculating the functions of DCI/EMI, and developing the response surface model, will be embedded in the process templates at different levels in the format of knowledge. By invoking and instantiating these modules, reusing and populating the generated information into the *Uncertainty* `Slots` property is possible. Additionally, in the robust design template reusing and populating *ResponseFunction*, i.e., determining the information instances of the response function by instantiating the designing of experiments or reusing empirical models, is also possible.

6.3.4 Test Example: Design of Hot Rod Rolling System

In this section, the utility of the robust decision process ontology is tested using a hot rod rolling (HRR) design example. This example involves integrated design of material, product, and associated manufacturing processes, requiring several hierarchical decisions to manage the uncertainties involved. The designer (decision-maker) aims to identify the processing paths and material structures that satisfy specific product and manufacturing process-level requirements and performances [52]. In [53, 54], the authors present model-based methods to realize engineered materials, products, and the associated manufacturing processes to link the material processing-structure–property-performance domains. In this work, we explain the robust decision process chain of hot rod rolling simulation models, and represent and capture the robust design uncertainty knowledge for the simulation models using the developed ontology.

(1) **Design Problem of an HRR System**

In a steel manufacturing process, rods and bars involve serial unit operations like continuous casting, reheating, rolling, and cooling. To integrate horizontal and vertical information flows for the HRR process chain problem, it is necessary to model the material behaviors at different scales and within different unit operations [55]. Developing steels with high-quality performance and a range of properties is vital for manufacturing designers. The steel products' mechanical properties and performance can be improved by manipulating the material processing and tailoring the microstructure of the steel material generated [29]. Nevertheless, this strategy involves a highly complex decision process chain. The designer has to deal with the constraints and bounds, conflicting goals, system variables, and process parameters, etc., because of a large amount of information for the sequential manufacturing process and material processing flows.

Traditional plant trials are usually expensive and time-consuming, and therefore, simulation-supported trials exploit the computational modeling at various scales to obtain the desired mechanical performance and properties for the steel products. The HRR simulation model chain adopts the method proposed by Nellippallil et al. [52, 55], who define the forward information flows of the hot rolling and cooling manufacturing stages, which are part of the hot rod rolling process. The quality of the end product rod is measured employing the identified mechanical property space, i.e., hardness (*HV*), tensile strength (*TS*), and yield strength (*YS*) that depend on the final microstructure after cooling. The rod's properties are related to the pearlite (X_p) and ferrite (X_f) fractions, the chemical composition of the material (e.g., silicon [Si], nitrogen [N], manganese [Mn], etc.), the pearlite interlamellar spacing (S_o) and the ferrite grain size after the austenite to ferrite and pearlite (FGS, D_α) transformation. Thus, the microstructure space is developed during the HRR process cooling state, where the inputs involve the chemical composition after the rolling stage process, the cooling rate (CR), and the final austenite grain size (AGS, D). The integrated hot rolling and cooling design process complete HRR's forward material workflow and establishes the material system's process-structure–property-performance hierarchy.

Several authors have highlighted the challenges related to the simulation-based multiscale material design [54, 56] that are related to (1) uncertain material models (including responses, parameters, input factors, etc.) due to lack or simplification/idealization of complete knowledge and (2) the uncertainty propagation due to Olson's relationship of processing-structure–property-performance or hierarchical information dependence in a multiscale model chain. To determine a solution having satisfactory robustness for the specific design requirements, integrating the horizontal and vertical HRR design simulation model chain of the problem affords the designer the opportunity to perform a robust decision-based design exploration of the manufacturing process chain. Hence, uncertainty management while designing a robust decision workflow is essential. This management process requires the designer to determine the necessary sequence of activities for the systematic design space exploration under uncertainty, ensuring the proper combination of the robust design information that meets the various constraints and the conflicting goals. To highlight uncertainty management during designing decision workflows, we demonstrate how a designer can capture, represent, and document the reusable information in the HRR problem and thus assist designers to explore the design space by exploiting the robust decision process template.

(2) Robust Decision Process for the Processing-Microstructure Simulation Models

In the HRR's cooling stage, the process designer must initially determine the basic features of the problem model before creating the processes to solve and explore the problem. Hence, the robust design template allows the designer to define the problem's initial robust design space, such as the process/product model, fixed parameters, response, and factors (noise, control, and signal factors). Fig. 6.15 presents a robust template instance for the cooling module as created based on Sect. 6.3.2 Step 1, displaying a graphical expression to design the robust design hierarchies. Furthermore, the integrated uncertainty management information and the related robust decision processes should also be captured and documented to assist the process designer in finding a robust solution. For the HRR cooling module, the *Robust_Template* `Instance` "HRR_RD_CoolingModule" is established and populates the corresponding `Slots` instances. In this work, we highlight the same by employing Fig. 6.15, where the designer adopts the empirical models to process a microstructure simulation chain [57], with the responses of pearlite interlamellar spacing S_o, ferrite grain size D_α, and ferrite fraction X_f defined as the control factors' function (inputs

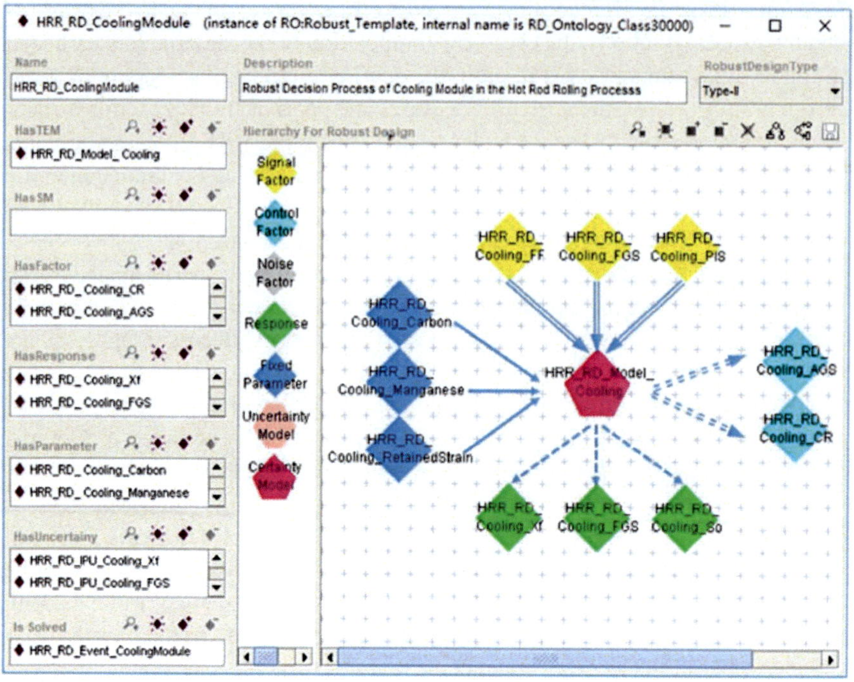

Fig. 6.15 Instance of the robust design template for cooling module in HRR process [51]

of microstructure space) and some related fixed parameters. Details are presented in the `Instance` of PEI-X process template "HRR_RD_Event_ CoolingModule" embedded in the robust design template.

Figure 6.16 shows that the hierarchical decision workflows are established based on Step 2 of the method presented in Sect. 6.3.2. Adopting the definition of the various process templates types of [41], the "HRR_RD_Event_CoolingModule" `Instance` is an "**Event**" process template. The primary function of an event process template is partitioning the design problem into some decision and associated task support problems, then constructing their execution sequences. An additional function is populating the design object, the design decision, task information, and the necessary knowledge to be reused in the design activities. For the cooling module case, the event instance inputs comprise the embedded "HRR_RD_Cooling_DesignRequirement" and "HRR_RD_Cooling_Design Space" `Instances`, and "HRR_RD_Cooling_Solution" is the output. The event support instance problem is split into three related tasks and one decision activity; namely, `Instances` "T2#1: ClarifyingTask", "T2#2: EstimatingUncertainty", "D1#1: compromiseDSP", and "T2#3: ExploringSolutionSpace" are created and populated. The reader is referred to [29] and [4] for further details on the procedure and slots per template instance. The information flows among those support problem entities which are represented through the `Instances` interface, e.g., "Interface: E1-T1#1",

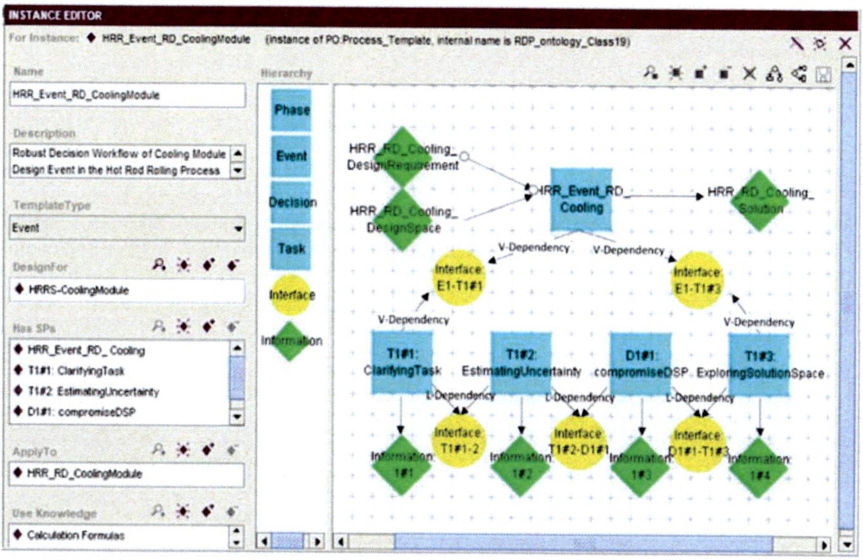

Fig. 6.16 Instance of the hierarchical decision workflow embedded in the robust template for cooling module [51]

"Interface: 1#1–2", etc., with a detailed description of the process template interface presented in [5].

The various activity modules in the process template aim to identify process-related attributes, with Table 6.10 presenting the primary information within the event process template instance. For example, the information instance considering the design space and requirements are populated to encapsulate and document the cooling module problem statement in the HRR process along with the relevant target and upper requirement limit (URL) for the signal factors. This information is essential for the designer to determine the basic problem/process model elements in "Task1#1", which is essential to instantiate the robust design template presented earlier. It is also essential to identify the management uncertainty type based on this information and calculate the related uncertainty in the follow-up tasks. Here, the "Task1#2" `Instance` follows the robust design Type II and the DCI is calculated to measure the system's robustness and performance by utilizing the variable response and the mean response functions.

We determine the uncertainty management module in the robust design template based on the previous attribute information and tasks. For instance, we populate the input parameter uncertainty management instances per response in the cooling module's product/process model. Window "①" in Fig. 6.17 illustrates that the "HRR_RD_Cooling_X_f" `Instance` is developed, while the variance response function instances and control factor deviation are shown in window "②" and "③", respectively. These template modules capture the necessary information of a specific design problem through the Slots for a robust design.

Table 6.10 Main attribute content of the information embedded in the event process template for cooling module [51]

Information	Main attribute content
HRR_PM_Solution space	● System constraints: DCI_{S_o}, DCI_{D_α}, DCI_{X_f} ● System goals: S_o, D_α, X_f ● System variables: *D*, *CR*
HRR_PM_Design requirement	● Achieving the lower value of the Microstructure Space (S_o, Pearlite Interlamellar Spacing, D_α, Ferrite Grain Size, and X_f, Ferrite Fraction) ● URL, $X_f = 0.75$; URL, $D_\alpha = 30$ μm; URL, $S_o = 0.2$ μm ● $\text{DCI}_{\text{target}}$, $X_f = 10$; $\text{DCI}_{\text{target}}$, $D_\alpha = 10$; $\text{DCI}_{\text{target}}$, $S_o = 10$
HRR_PM_Design space	● x_1, Cooling Rate (*CR*) ● x_2, Austenite Grain Size (*D*) | ● $x_1 = [11, 100]$ (K/min) ● $x_2 = [30, 100]$ (μm) | ● $\Delta x_1 = \pm 10$ (K/min) ● $\Delta x_2 = \pm 10$ (μm)
Information2#1	● Factor (control factors: *CR*, *D*; signal factors: X_f, D_α, S_o) ● Response ● Variation (deviation: Δx) ● Fixed parameter ([C], [Mn], ε_r) ● Response model
Information2#2	● The Type of Robust Design (Type II) ● Response Functions (mean response function and response variation unction) ● The DCI for each response
Information2#3	● The results of robust cDSP model with different design preference
Information2#4	● The satisfying robust solution

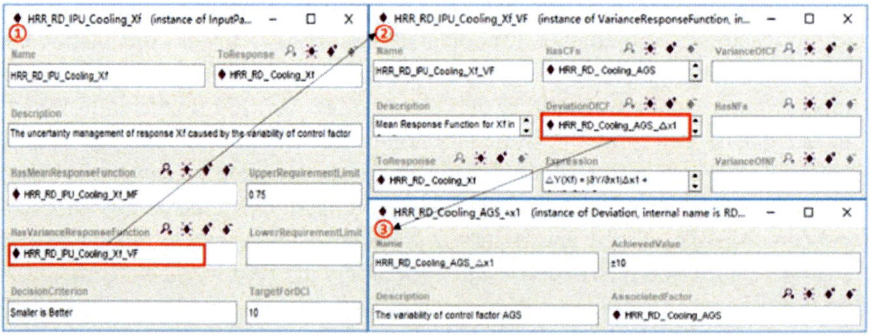

Fig. 6.17 Instances of the Uncertainty Management (IPU) embedded in the robust template for cooling module [51]

Uncertainty management aims to assist the designer in determining the optimum robust design information combination, which meets the various constraints and conflicting goals. Therefore, the cDSP template for microstructure space ("D1#1 `Instance`: compromise DSP") is formulated with DCI goals that capture the microstructure requirements under uncertainty [4] and [57]. Based on the results

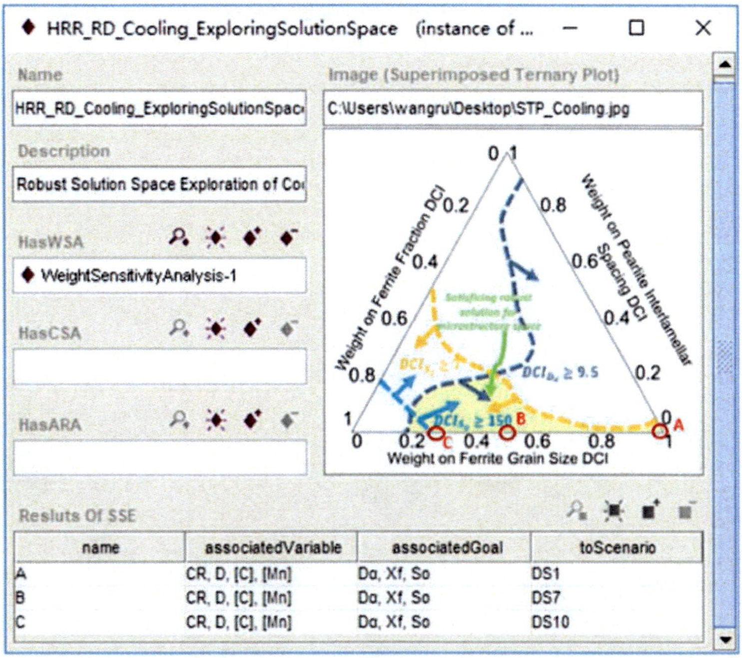

Fig. 6.18 Task instance of the solution space exploration for cooling module [51]

of the cDSP-DCI model, the designer can explore the solution space with different design preferences in the task `Instance` "T1#3: ExploringSolution Space" presented in Fig. 6.17. The weight sensitivity analysis within the solution space exploration instance is illustrated in Fig. 6.18. The superimposed ternary plot (microstructure solution space) and the solution spaces of interest determine the robust solution regions while considering multiple conflicting goals. For the cooling module, the designer determines a region with $DCI_{S_o} \geq 150$, $DCI_{D_\alpha} \geq 9.5$, and$DCI_{X_f} \geq 7$, based on the requirements of each decision criterion and goal. To further analyze and assist decision-making, we choose from the region identified three solution points (A, B, C) within the optimum region satisfying all the robust design goals. Ref. [8] introduces the details of the solution space exploration template.

(3) **Robust Decision Process for the Microstructure-Mechanical Simulation Models**

We create a robust design template appropriate for the HRR cooling by exploiting the uncertainty management and the embedded PEI-X process templates combined with reusing the robust decision workflows information. For this scenario, empirical models are utilized to define the employed response functions. The empirical models are commonly response surface models developed through the design of the involving experiment, in some sense, the response model uncertainties. Nevertheless, the process designer has adequate confidence to ignore this uncertainty impact

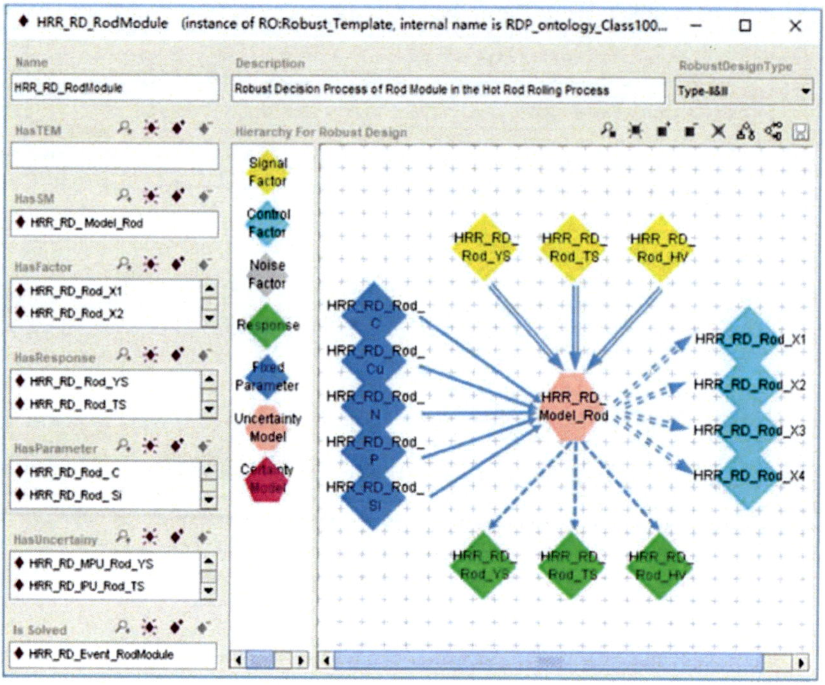

Fig. 6.19 Instance of the robust template for rod module in HRR process [51]

and only consider the control factors variations (IPU). Another scenario involves the uncertainty derived from the response model by instantiating the HRR's rod module robust design template in the process chain (Fig. 6.19).

In the HRR problem examined here, next to the microstructure correlation calculation (cooling module) is the rod module that predicts the mechanical properties of hardness (HV), tensile strength (TS), and yield strength (YS). Table 6.11 presents the design requirements for the rod module, from which the essential robust design elements such as fixed parameters, responses, signal factors, and control factors are populated. The difference from the previous case is that the process/product model `Instance` "HRR_RD_Model_Rod" is established in the robust design template instance by considering the response function uncertainty. The designer organizes these robust design elements using a graphical hierarchy scheme and identifies the robust design type according to each element's attribute information. Type II and Type III robust design types exist for the rod module case due to uncertainty of input and model parameters (i.e., IPU and MPU). Additionally, the hierarchical decision workflows are established to address the corresponding robust design, where the designer can reuse the cooling module's PEI-X process template.

The "HRR_RD_MPU_Rod_YS" `Instance` manages the model parameter uncertainty for the response yield strength (YS). During the uncertainty task estimation through this instance, the related sequential computability routines based on Step 3 are executed to create and populate the detailed information attributes. Figure 6.20

Table 6.11 Main information embedded in the robust design template for rod module [51]

Item	Main Information		
Design requirement	● Achieving the larger value of the mechanical property space (*YS*, Yield Strength; *TS*, Tensile Strength; *HV*, Hardness) ● $LRL_{YS} = 200$ MPa; $LRL_{TS} = 450$ MPa; $LRL_{HV} = 130$ ● $EMI_{target,}$ $YS = 3$; $DCI_{target,}$ $TS = 8$; $DCI_{target,}$ $HV = 8$		
Control factors	● x_1, Ferrite Grain Size (D_α) ● x_2, Ferrite Fraction (X_f) ● x_3, Pearlite Interlamellar Spacing (S_o) ● x_4, Manganese concentration after ooling ([Mn])	● $x_1 = [5, 25]$ (μm) ● $x_2 = [0.1, 1]$ ● $x_3 = [0.15, 0.25]$ (μm) $x_4 = [0.7, 1.5]$ (%)	● $\Delta x_1 = \pm 3$ (μm) ● $\Delta x_2 = \pm 0.1$ ● $\Delta x_3 = \pm 0.01$ (μm) ● $\Delta x_4 = \pm 0.01$
cDSP-EMI/DCI	● System variables: D_α, X_f, S_o, [Mn] ● System goals: Maximize EMI for *YS*, Maximize DCI for T*S*, Maximize DCI for *HV* ● System constraints: $EMI_{YS} \geq 1$, $DCI_{TS} \geq 1$,$DCI_{HV} \geq 1$		

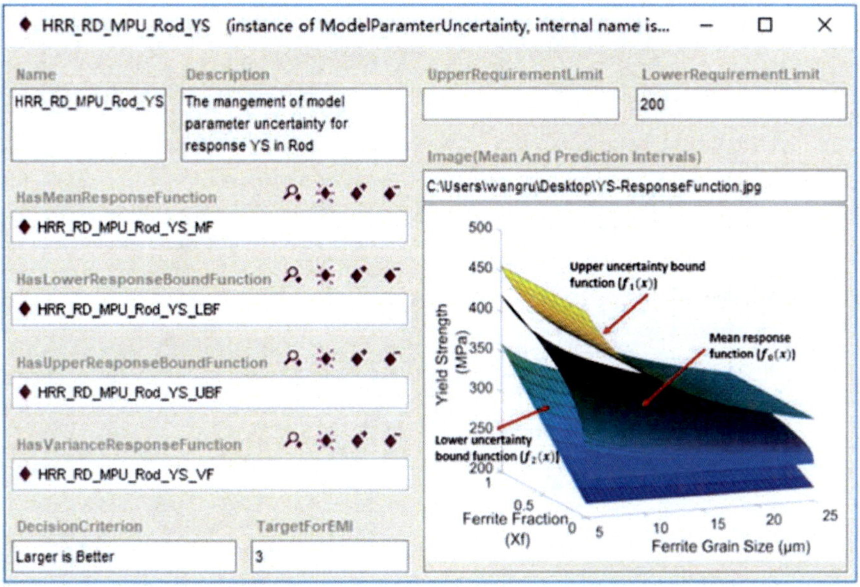

Fig. 6.20 Instance of the Uncertainty Management (MPU) embedded in the robust template for cooling module [51]

depicts the model response (YS) variation in terms of the maximum/minimum prediction intervals functions (i.e., lower/upper uncertainty bound function) and the mean response function. Utilizing these mathematical functions, the designer can estimate the lower and upper response deviations and then calculate the EMI based on the decision criterion. For the rod module simulation model case, the EMI/DCI decision criterion is "Larger is Better", highlighting that the mean responses, i.e., *YS*, *TS*, and *HV*, must be further away from the lower requirement limit (LRL) defined and close to the EMI/DCI defined target. The cDSP's formulation objective for the EMI/DCI is finding the control factors' mean value (x_1~x_4) satisfying the given bounds and performance goals.

Solving the cDSP-EMI/DCI model under various design preferences enables the designer to obtain the system goals values under the identified uncertainties: DCI_{HV}, DCI_{TS}, and EMI_{YS}. To attain high *DCI* and *EMI* values for the *YS*, *TS*, and *HV* model response, the designer must define satisfying robust solution regions for the several conflicting goals. Figure 6.21 identifies in the task instance of the solution space a joint robust solution region (light-yellow region) with $DCI_{HV} \geq 7$, $DCI_{TS} \geq 6$, and $EMI_{YS} \geq 1.5$ that complies with the robust design requirements related to the conflicting mechanical property goals. To assist the designer's

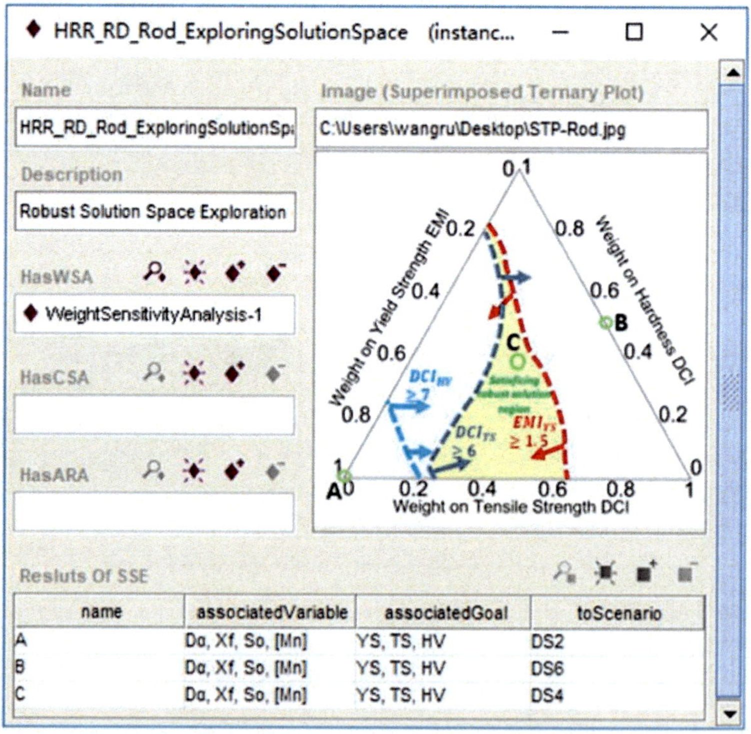

Fig. 6.21 Task instance of the solution space exploration for rod module [51]

decision-making process, we investigate three solution points (A, B, C). Point A is the most robust region for *YS* managing a high *EMI* but is not appealing regarding *TS* and *HV* posing low *DCIs*. Similarly, point B has a high DCI_{TS} affording the most robust *TS* region but presents the lowest *YS* with a very low EMI. In contrast, point C lying inside the satisfying robust solution regions attains the highest DCI_{HV} and simultaneously poses the most robust *HV* goal region satisfying the robust design requirements of the remaining goals. Therefore, point C is the optimum choice for the subsequent manufacturing operations.

6.4 Empirical Structural Validity

Empirical structural validation is to build confidence in the appropriateness of the test example problems chosen for illustrating and verifying the performance of the framework and methods. Model-based realization of complex engineered systems involves managing information associated with models that are typically incomplete, inaccurate, and not of equal fidelity. Designing such systems, therefore, demands the designers to carry out rapid and systematic exploration of design space to identify solutions that are relatively insensitive to the associated uncertainties. It is always essential but difficult to represent and capture the uncertainty knowledge, some practical limitations have also been realized with regard to industrial implementations, especially in the context of knowledge-intensive complex engineered systems. In view of DBD, system design involves a sequence of decision-making, that is decision workflows, which require a combination of analytical models and a way to synthesize the information generated by decision models. Therefore, a template-based ontological method for design space exploration and uncertainty management is introduced that supports the designing of decision workflows to ensure decision-making with the features of robustness, feasibility, and comprehensiveness, taking into account the goals of enhancing the design automation and intelligence in designing of decision workflows.

6.5 Where We Are and What Comes Next?

In this chapter, we develop two ontologies for representing the knowledge related to design space exploration and uncertainty management which serve as the foundation for supporting knowledge-based design space exploration on the PDSIDES platform. In the next chapter, we will introduce the Cloud-Based PDSIDES (CB-PDSIDES).

References

1. Panchal, J. H., Fernandez, M. G., Paredis, C. J. J., & Mistree, F. (2012). Reusable design process modeling via modular, executable, decision-centric templates. In *Proceedings of the 10th AIAA/ISSMO Multidisciplinary Analysis and Optimization Conference.* Paper MAO-49. https://doi.org/10.2514/6.2004-4601.
2. Panchal, J. H., Fernandez, M. G., Paredis, C. J. J., Allen, J. K., & Mistree, F. (2009). A modular decision-centric approach for reusable resign rrocesses. *Concurrent Engineering, 17*(1), 5–19.
3. Schönberg, F., & Messer, M. (2018). Decision data model in virtual product development. *Computers & Industrial Engineering, 22,* 106–124.
4. Ming, Z., Yan, Y., Wang, G., Panchal, J. H., Goh, C.-H., Allen, J. K., & Mistree, F. (2016). Ontology-based executable design decision template representation and reuse. *Artificial Intelligence in Engineering Design and Manufacturing (AIEDAM), 30*(4), 390–405.
5. Ming, Z., Wang, G., Yan, Y., Dal Santo, J., Allen, J. K., & Mistree, F. (2017). An ontology for reusable and executable decision templates. *Journal of Computing and Information Science in Engineering, 17*(3), 031008.
6. Noy, N. F., & McGuinness, D. L. (2001). Ontology development 101: A guide to creating your first ontology. In *Stanford knowledge systems laboratory technical report KSL-01-05 and Stanford medical informatics technical report SMI-2001-0880*, Stanford, CA.
7. Torga, M. Å., Andreasen, M. M., & Marjanoviä, D. (2010). The design ontology: Foundation for the design knowledge exchange and management. *Journal of Engineering Design, 21*(4), 427–454.
8. Wang, R., Wang, G., Yan, Y., Chen, S., Allen, J. K., & Mistree, F. (2017). A framework for knowledge-intensive design decision support in model based realization of complex engineered systems. *Proceedings IEEE International Conference on Industrial Engineering and Engineering Management*, pp. 230–234.
9. Moon, S. K., Simpson, T. W., Shu, J., & Kumara, S. R. T. (2009). Service representation for capturing and reusing design knowledge in product and service families using object-oriented concepts and an ontology. *Journal of Engineering Design, 20*(4), 413–431.
10. Rockwell, J., Grosse, I. R., Krishnamurty, S., & Wileden, J. C. (2009). A decision support ontology for collaborative decision making in engineering design. In *Proceedings of the 2009 International Symposium on Collaborative Technologies and Systems* (pp. 1–9), IEEE.
11. Noor, M. H. M., Salcic, Z., Kevin, I., & Wang, K. (2016). Enhancing ontological reasoning with uncertainty handling for activity recognition. *Knowledge-Based Systems, 114,* 47–60.
12. Yang, Y., & Calmet, J. (2005). OntoBayes: An ontology-driven uncertainty model. In *Proceedings Computational Intelligence for Modelling, Control, and Automation, 2005 and International Conference on Intelligent Agents, Web Technologies and Internet Commerce* pp. 457–463.
13. Ding, Z., Peng, Y., & Pan, R. (2005). BayesOWL: Uncertainty modeling in semantic web ontologies. *Studies in Fuzziness and Soft Computing, 204,* 3–29.
14. Costa, P. C. G. D., Laskey, K. B., & Laskey, K. J. (2005). PR-OWL: A Bayesian ontology language for the semantic web. *Proceedings of the Uncertainty Reasoning for the Semantic Web I, ISWC International Workshops, URSW 2005–2007, Revised Selected and Invited Papers* pp. 88–107.
15. Fenz, S. (2012). An ontology-based approach for constructing Bayesian networks. *Data & Knowledge Engineering, 73*(2), 73–88.
16. Lim, S. C. J., Ying, L., & Han, T. L. (2012). An exploratory study of ontology-based platform analysis under user preference uncertainty. In *Proceeding ASME 2012 International Design Engineering Technical Conferences and Computers and Information in Engineering Conference*, pp. 519–528.
17. Costa, P. C. G., Laskey, K. B., Blasch, E., & Jousselme, A. L. (2012). Towards unbiased evaluation of uncertainty reasoning: The URREF ontology. In *Proceedings International Conference on Information Fusion*, pp. 2301–2308.

18. Sim, S. K., & Duffy, A. H. (2003). Towards an ontology XE "ontology" of generic engineering design activities. *Research in Engineering Design, 14*(4), 200–223.
19. Singh, S., Ghosh, S., Jayaram, J., & Tiwari, M. K. (2019). Enhancing supply chain resilience using ontology-based decision support system. *International Journal of Computer Integrated Manufacturing, 32*(7), 642–657.
20. Kang, E., Jackson, E., & Schulte, W. (2010). An approach for effective design space exploration. In *Proceedings of the Monterey Workshop* (pp. 33–54), Springer.
21. Chong, Y. T., Chen, C. H., & Leong, K. F. (2009). A heuristic-based approach to conceptual design. *Research in Engineering Design, 20*(2), 97–116.
22. Smith, W. F., Milisavljevic, J., Sabeghi, M., Allen, J. K., & Mistree, F. (2015). The realization of engineered systems with considerations of complexity. *International Design Engineering Technical Conferences and Computers and Information in Engineering Conference,* Paper Number DETC2015-46211.
23. Bloebaum, C. L., & McGowan, A.-M. R. (2010). Design of complex engineered systems. *Journal of Mechanical Design, 132*(12), p. 120301.
24. Tribus, M. (2016). Rational descriptions, decisions, and designs: Pergamon Unified Engineering Series, Elsevier.
25. Estefan, J. A. (2007). Survey of model-based systems engineering (MBSE) methodologies. *INCOSE MBSE Focus Group, 25*(8).
26. Micouin, P. (2014). Model based systems engineering: fundamentals and methods, John Wiley & Sons.
27. Chen, W., Allen, J. K., Mavris, D. N., & Mistree, F. (1996). A concept exploration method for determining robust top-level specifications. *Engineering Optimization+ A35, 26*(2), 137–158.
28. Nellippallil, A. B., Allen, J. K., Mistree, F., Vignesh, R., Gautham, B. P., & Singh, A. K. (2017). Agoal-oriented, inverse decision-based design method to achieve the vertical and horizontal integration of models in a hot-rod rolling process chain. *ASME Design Automation Conference,* Paper Number DETC2017-67570, Cleveland, Ohio, USA.
29. Wang, R., Nellippallil, A. B., Wang, G., Yan, Y., Allen, J. K., & Mistree, F. (2018). Systematic design space exploration using a template-based ontological method. *Advanced Engineering Informatics, 36*, 163–177.
30. Hoyle, C. J., & Chen, W. (2009). Product attribute function deployment (PAFD) for decision-based conceptual design. *IEEE Transactions on Engineering Management, 56*(2), 271–284.
31. Mistree, F., Bras, B., Smith, W. F., & Allen, J. K. (1995). Modelling design processes: A conceptual, decision-based perspective. *Engineering Design and Automation 1*(4), 209–221.
32. Simpson, T. W., Peplinski, J., Koch, P. N., & Allen, J. K. (2001). Metamodels for computer-based engineering design: Survey and recommendations. *Engineering with Computers, 17*(2), 129–150.
33. Montgomery, D. C., & Myers, R. H. (1995). *Response surface methodology: process and product optimization using designed experiments.* Wiley-Interscience.
34. Bras, B., & Mistree, F. (1993). Robust design using compromise decision support problems. *Engineering Optimization, 21*(3), 213–239.
35. Mistree, F., Hughes, O. F., & Bras, B. (1993). Compromise decision support problem and the adaptive linear programming algorithm. *Progress in Astronautics and Aeronautics, 150*, 251–251.
36. Nellippallil, A. B., Song, K. N., Goh, C.-H., Zagade, P., Gautham, B., Allen, J. K., & Mistree, F. (2017). A goal-oriented, sequential, inverse design method for the horizontal integration of a multi-stage hot rod rolling system. *Journal of Mechanical Design, 139*(3), 031403.
37. Sabeghi, M., Smith, W., Allen, J. K., & Mistree, F. (2016). Solution space exploration in model-based realization of engineered systems. In *Proceeding ASME 2015 International Design Engineering Technical Conferences and Computers and Information in Engineering Conference,* Paper Number DETC2016-59399.
38. Nellippallil, A. B., Vignesh, R., Allen, J. K., Mistree, F., Gautham, B. P., & Singh, A. K. (2017). A decision-based design method to explore the solution space for microstructure after cooling stage to realize the end mechanical properties of hot rolled product. In *4th World Congress on Integrated Computational Materials Engineering (ICME 2017)* Ypsilanti, Michigan, USA.

39. Jones, S., & Bhadeshia, H. (1997). Kinetics of the simultaneous decomposition of austenite into several transformation products. *Acta Materialia, 45*(7), 2911–2920.
40. Robson, J., & Bhadeshia, H. (1997). Modelling precipitation sequences in power plant steels Part 1–Kinetic theory. *Materials Science and Technology, 13*(8), 631–639.
41. Wang, R., Wang, G., Yan, Y., Sabeghi, M., Ming, Z., Allen, J. K., & Mistree, F. (2017). Ontology-based representation of meta-design in designing decision workflows. In *ASME 2017 International Design Engineering Technical Conferences and Computers and Information in Engineering Conference*, Paper Number DETC2017-67817.
42. Simon, H. A. (1996). The sciences of the artificial, MIT press.
43. Panchal, J. H., Choi, H.-J., Allen, J. K., McDowell, D. L., & Mistree, F. (2007). A systems-based approach for integrated design of materials, products and design process chains. *Journal of Computer-Aided Materials Design, 14*(1), 265–293.
44. Phadke, M. S. (1989). Quality engineering using robust design, Prentice-Hall International.
45. Choi, H. J. (2005). *A robust design method for model and propagated uncertainty*. Ph.D. dissertation, Georgia Institute of Technology.
46. Chen, W., Allen, J. K., Tsui, K. L., & Mistree, F. (1996). A procedure for robust design: minimizing variations caused by noise factors and control factors. *Journal of Mechanical Design, 118*(4), 478–485.
47. Simpson, T. W., Chen, W., Allen, J. K., & Mistree, F. (1997). Designing ranged sets of top-level design specifications for a family of aircraft: application of design capability indices. In *Proceeding SAE World Aviation Congress and Exhibition,* Paper Number 975513.
48. Choi, H. J., Austin, R., Allen, J. K., Mcdowell, D. L., Mistree, F., & Benson, D. J. (2005). An approach for robust resign of reactive power metal mixtures based on non-deterministic micro-scale shock simulation. *Journal of Computer-Aided Materials Design, 12*(1), 57–85.
49. Choi, H., McDowell, D. L., Allen, J. K., Rosen, D., & Mistree, F. (2008). An inductive design exploration method for robust multiscale materials design. *Journal of Mechanical Design, 130*(3), p. 031402.
50. Garcíapeñalvo, F. J., Pablos, P. O. D., García, J., & Therón, R. (2014). Using OWL-VisMod through a decision-making process for reusing OWL ontologies. *Behaviour & Information Technology, 33*(5), 426–442.
51. Wang, R., Nellippallil, A. B., Wang, G., Yan, Y., Allen, J. K., & Mistree, F. (2019). Ontology-based uncertainty management approach in design of rubust decision workflows. *Journal of Engineering Design, 30*(10–12), 726–757.
52. Nellippallil, A. B., Rangaraj, V., Gautham, B., Singh, A. K., Allen, J. K., & Mistree, F. (2018). An inverse, decision-based design method for integrated design exploration of materials, products, and manufacturing processes. *Journal of Mechanical Design, 140*(11), p. 111403.
53. Olson, G. B. (1997). Computational design of hierarchically structured materials. *Science, 277*(5330), 1237–1242.
54. Sinha, A., Bera, N., Allen, J. K., Panchal, J. H., & Mistree, F. (2011). Uncertainty management in the design of multiscale systems. *Journal of Mechanical Design, 135*(1), 011008.
55. Nellippallil, A. B., Song, K. N., Goh, C.-H., Zagade, P., Gautham, B., Allen, J. K., & Mistree, F. (2017). A goal-oriented, sequential, inverse design method for the horizontal integration of a multistage hot rod rolling system. *Journal of Mechanical Design, 139*(3), 031403.
56. McDowell, D. L. (2018). Microstructure-sensitive computational structure-property relations in materials design. In D. Shin, J. Saal (Eds.), *Computational Materials System Design* (pp. 1–25), Springer.
57. Nellippallil, A. B., Mohan, P., Allen, J. K., & Mistree, F., (2018). Robust concept exploration of materials, products and associated manufacturing processes. In *Proceeding ASME 2018 International Design Engineering Technical Conferences and Computers and Information in Engineering Conference, American Society of Mechanical Engineers*, Paper Number DETC2018-85913.

Chapter 7
Extending PDSIDES to CB-PDSIDES: New Opportunities in Design Engineering 4.0

In this chapter, we revisit the problem, summarize the approach to address the problem, review the validation strategy, and identify the main benefits and contributions of this monograph. Based on the summary of this monograph, we speculate about the opportunities for improving PDSIDES by proposing a framework for a Cloud-Based Platform for Decision Support in the Design of engineered Systems (CB-PDSIDES) and identify several challenges and research questions for the realization of CB-PDSIDES. Finally, we end the monograph with some closing comments. Table 7.1 is a summary of the topics discussed in previous chapters. A detailed summary of the monograph in terms of problem, approach, validation, benefits/contributions is presented in Sect. 7.1.

7.1 Summary of Monograph

Problem: A design revolution is underway, where the focus is to adopt a "human-cyber-physical view of the systems realization ecosystem" to accommodate customers' personalized demands as a result of the technological advances of Industry 4.0 and develop proper fulfillment capabilities to respond to dynamic markets in the new era. This revolution is called "Design Engineering 4.0". The human dimension of Design Engineering 4.0 not only emphasize the importance of the experience of customers at the front-end, but also the experience of all the stakeholders involved in the product realization process at the back-end, especially designers who are the decision-makers in the design of products. From a Decision-Based Design perspective, we believe that the experience of designers can be augmented by supporting them to make *rapid*, *informed*, *satisficing*, *robust*, and *visualized* decisions in the age of

© The Author(s), under exclusive license to Springer Nature Switzerland AG 2022

Z. Ming et al., *Architecting A Knowledge-Based Platform for Design Engineering 4.0*,
https://doi.org/10.1007/978-3-030-90521-7_7

Table 7.1 Topics of previous chapters

Architecture constituents	Summary interpretation for the platform (Chaps. 1 and 2)	Chapter 3	Chapter 4	Chapter 5	Chapter 6
Elements	What?	What?	What?	What?	What?
Components	Foundational concepts, techniques for architecting a knowledge-based decision support platform	Ontologies for individual decisions, including selection, compromise, and coupled decisions	The functionalities (components) of the PDSIDES platform	Ontology for meta-design of designing decision workflows	Ontology for robust design space exploration
Connectors	The media that links the components, namely, the platform	Ontology	The PDSIDES platform	Ontology	Ontology
Form	How?	How?	How?	How?	How?
Component roles	Different roles that the components play in the platform	Knowledge models of individual decisions	Implementing and testing the utility of PDSIDES functionalities	Knowledge model of decision-based design processes	Knowledge model of uncertainty management in decision-based design
Properties	Attributes or features of the components	Requirements for developing ontologies for DSP templates	Requirements of developing PDSIDES	Requirements of developing the meta-design ontology	Requirements of developing the robust design ontology
Relationship	The relationships among different components in the platform	The DSPT framework	Original design—adaptive design—variant design	The PEI-X diagram	The robust design framework
Rationale	Why?	Why?	Why?	Why?	Why?
Motivation	The underlying reason to use this platform, driven by requirements	To build a knowledge based for decision-based design	To prototype PDSIDES	To build a knowledge based for decision-based meta-design	To build a knowledge based for decision-based design under uncertainty

(continued)

Table 7.1 (continued)

Architecture constituents	Summary interpretation for the platform (Chaps. 1 and 2)	Chapter 3	Chapter 4	Chapter 5	Chapter 6
Assumptions	Generalizations or simplifications used to guide the selection of elements	There are only two primary types of decisions in design: selection and compromise	There are three types of users: template Creators, Editors, and Implementers	Any design process can be represented using phases, events, and information	There are four types of uncertainties in design
Interpretation	Understanding how the platform is composed to deliver decision support	Understanding how individual decision-related knowledge is represented using ontology	Understanding how PDSIDES is realized to support designers making decisions in the design of real-world engineered systems	Understanding how meta-design-related knowledge is represented using ontology	Understanding how robust design-related knowledge is represented using ontology

Industry 4.0. Accordingly, this monograph is anchored in architecting a ***knowledge-based platform*** for providing decision support with the aforementioned characteristics in the design of engineered systems. Major challenges in establishing the architecture of such a platform are: (1) the management of multidisciplinary, domain-dependent and -independent, heterogeneous knowledge and its reuse in supporting different types of decisions, (2) the formulation of individual decisions and decision workflows, (3) the exploration of a discrete and/or continuous multidimensional solution space, (4) the management of different types of uncertainty from multiple sources, and (5) the offer of user/activity specific decision support.

Approach: In this monograph, we describe a knowledge-based computational platform, PDSIDES, to support human designers making decisions that have a major effect on closing design freedom associated with the realization of complex engineered systems. The architecture of PDSIDES is an integration of the constructs from Decision-Based Design (e.g., cDSP, sDSP, and PEI-X, etc.) and Knowledge-Based Engineering (i.e., templates and ontologies). The architecting process of PDSIDES is described as follows. We view engineering design as a decision-making process and use sDSP and cDSP as the constructs for formulating the two primary types of decisions, namely, selection and compromise in design. These two constructs are the building blocks for representing any complex design. Based on sDSP and cDSP, we develop modular templates to enable the reusability and executability of decision-related knowledge in a computational environment, and use ontologies to formally represent the templates and the associated modules so as to facilitate knowledge sharing and exchange. Based on the constructs including sDSP, cDSP, the coupling of both, and the associated templates and ontologies, we implement a prototype of PDSIDES that can support three types of users—template creators (experts), template editors (senior designers), template implementers (novice designer), in different types of activities including original design, adaptive design, and variant design. In order to extend the functionalities of PDSIDES to support meta-design of decision workflows and robust design space exploration, we create an ontology based on the PEI-X construct which is an icon-based object-oriented diagram for modeling the hierarchy of the design process of complex systems. Both the partitioning and the integration characteristics of the complex workflow hierarchy are captured using the PEI-X diagram. In order to manage the uncertainty pertaining to decision workflows in PDSIDES, we create an ontology for robust design solution space exploration which represents the knowledge related to a formal procedure for identification of satisficing solutions that are relatively insensitive to variations.

Validation: The architecture of the knowledge-based platform for decision support in the design of engineered systems is validated using the validation-square approach that consists of both theoretical and empirical validation. Theoretical validation is carried out by justifying the structural correctness of the architecture and its generality. Empirical validation is carried out using several examples including light switch cover material selection, pressure vessel design, hot rod rolling system design, and heat exchanger design problems. Specific example problems are used to test the utility of the components as well as the overall performance of the platform. Since the constructs used to establish PDSIDES such as cDSP, sDSP, PEI-X, and ontologies are domain-independent, PDSIDES has the generality to support decisions in other design problems beyond the examples considered in this monograph. Applications of PDSIDES can extend to other industries such as energy (e.g., the design of power grids), healthcare (e.g., the design of a hospital system), supply chains (e.g., the design of a supply network), sports (e.g., the design of an American football helmet), etc., where design plays a critical role and decision support is needed.

Main benefits/Contributions: The main benefits/contributions from the monograph are anchored in the theory fundamentals and tools for architecting a knowledge-based platform for decision support in the realization of complex engineered systems. The theory fundamentals and tools include constructs for formulating design decisions, knowledge modeling schemes, diagrams for designing decision workflows, and some analytical methods for robust design space exploration under uncertainty. With these theory fundamentals and tools being organized and integrated into a computational platform, designers in the Industry 4.0 era will be able to make more *rapid*, *informed*, *satisficing*, *robust*, and *visualized* decisions in the design of engineered systems and the associated design processes. To the best of our knowledge, this is the first monograph that provides a systematic architecture for knowledge-based decision support in the design of engineered systems. This monograph will serve as an important guide for both students as well as practitioners who are interested in architecting computational platforms for decision support in the design of complex engineered systems.

The relationship of research efforts with the constructs of the architecture and connection between chapters of the monograph are shown in Fig. 7.1. The ontologies for multi-attribute selection decisions, multi-objective compromise decisions,

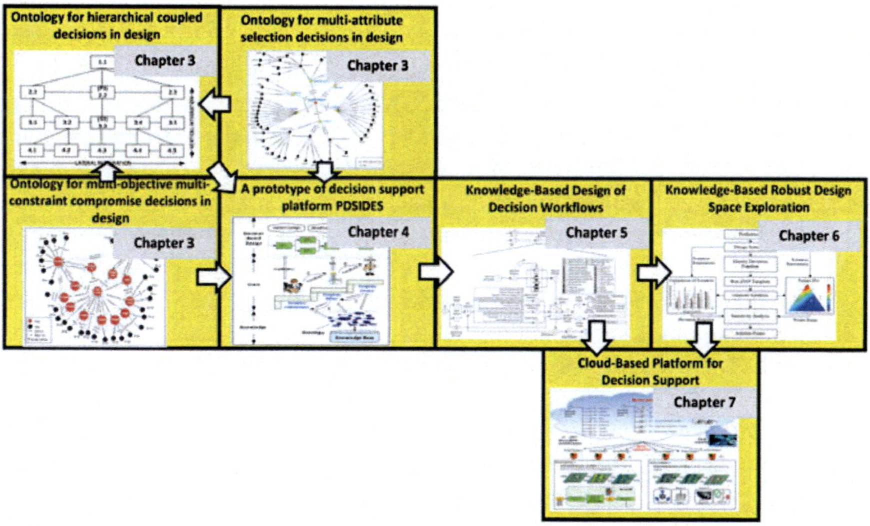

Fig. 7.1 Relationship of research efforts with the constructs of the architecture and connections among chapters of the monograph

and hierarchical coupled decisions are developed in Chap. 3. The three ontologies provide the knowledge representation schemes for the knowledge base of platform PDSIDES developed in Chap. 4. In Chap. 5, we extend the ontology to represent the knowledge of decision workflows based on the PEI-X diagram and enable the functionality of PDSIDES to support meta-design of decision workflows. The idea of representing phase and event-based procedural knowledge is extended from Chaps. 5 and 6 for representing robust design space exploration processes. The issues of uncertainty management and robust decision-making are addressed using PDSIDES in Chap. 6. Based on the current functionalities of PDSIDES, our focus in this chapter is on furthering the research vision by exploring opportunities for a service-oriented architecture for decision support using the cloud.

Given the contributions that have been made in the previous chapters of the monograph, we speculate about the opportunities for improving PDSIDES by proposing a framework for a Cloud-Based Platform for Decision Support in the Design of engineered Systems (CB-PDSIDES), which will be discussed in Sect. 7.2.

7.2 Cloud-Based Decision Support: Framework and Open Questions

Cloud-Based Design and Manufacturing (CBDM) is listed as one of the key features of Industry 4.0 by Thames and Schaefer [1] where they attempt to define the drivers of Industry 4.0 and the resulting innovations. According to Thames and Schaefer [1], CBDM refers to a service-oriented networked product development model in which service consumers are enabled to configure, select, and utilize customized product realization services ranging from computer-aided engineering software to reconfigurable manufacturing systems. The core of CBDM is the Service-Oriented Architecture (SOA) that enables the integration of heterogeneous design and manufacturing resources through services and provides interoperability between different platforms of service consumers and providers. The benefits of SOA include ubiquities access to design and manufacturing Resources, on-demand scalability and multi-tenancy, increased resource utilization, reduced capital cost and complexity, reduced maintenance cost accelerated time-to-market, attractive pay-as-you-go pricing, and democratization of innovation, etc. In the context of CBDM, we propose a SOA-based conceptual framework CB-PDSIDES which integrates PDSIDES with the cloud for future exploration, as shown in Fig. 7.2.

In CB-PDSIDES, decision-related knowledge is servitized and deployed in a service pool in the cloud, including declarative knowledge service (which is domain-dependent and used for specifying the required information such as alternatives, attributes, constraints, and goals, etc. to make a decision in a specific domain or problem), procedural knowledge service (which is domain-independent and provides the constructs such as the sDSP template and cDSP template for formulating decisions), and integrated knowledge services (which is the combination of declarative

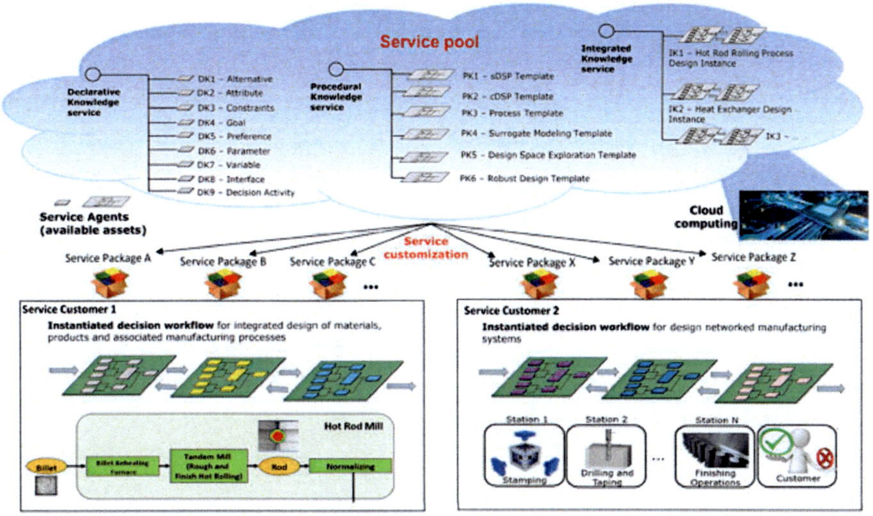

Fig. 7.2 A conceptual framework of CB-PDSIDES

and procedural knowledge and represents the "best-practice" of solving a design problem such as a hot rod rolling process design problem or a heat exchanger design problem). These services are uploaded to the cloud by service providers and form the available assets which can be customized by service customers to support their decisions in particular design problems. Service customers select and configure the services from the service pool and compose service packages according to the feature of their problems. For example, a service customer who wants to design a hot rod rolling system may select an existing instance from the integrated knowledge service category and then reconfigure it or select some entities from the declarative and procedural knowledge service category then compose these entities into an integrated service package for solving the problem. CB-PDSIDES provide the computing environment (in the air) for the service packages to execute and generate results for supporting customers' decisions.

7.2.1 Architecture of Cloud-Based PDSIDES

In Fig. 7.3, we show the architecture of CB-PDSIDES. The computing architecture for CB-PDSIDES follows the architecture of cloud-based design and manufacturing systems proposed by Wu and co-authors [2]. The architecture of CB-PDSIDES includes five layers {Nellippallil, 2019 #462}: (i) user layer, (ii) web portal layer, (iii) logic layer, (iv) virtual layer, and (v) physical layer.

Since CB-PDSIDES is deployed in the cloud, users can access CB-PDSIDES via PCs and smartphones over the internet. The web portal layer of CB-PDSIDES includes the user interaction GUI for accessing the design templates available. The

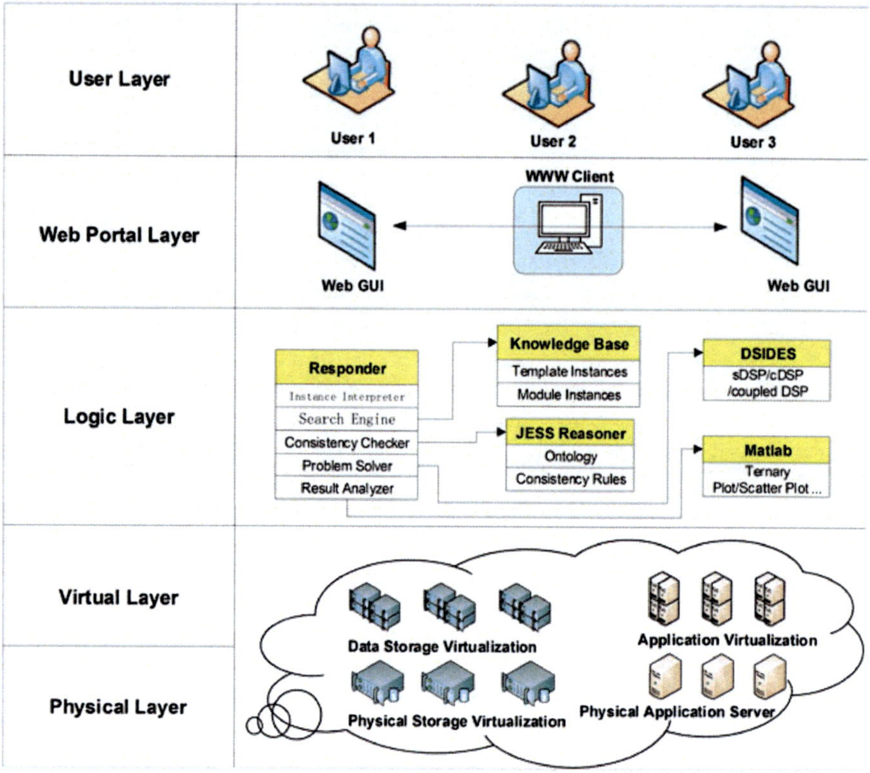

Fig. 7.3 Computing architecture of CB-PDSIDES [3]

user interaction GUI includes: the template searching and browsing GUI, designed for locating the required DSP templates and presenting them, the template creating and editing GUI designed based on the DSP template structures for instantiation and modification of the DSP templates, the template execution and analysis GUI designed for executing DSP templates and performing post-solution analysis. The GUI is allowed to communicate with the logic layer of CB-PDSIDES by a request-response mode using the HyperText Transfer Protocol (HTTP). The logic layer of CB-PDSIDES includes five main parts, namely, Response Server, Knowledge Base, JESS Reasoner, DSIDES, and a link to the commercial software MATLAB. The Response Server is the central "brain" that integrates the other four parts for responding to requests. The Response Server itself has five components: a search engine, an instance interpreter, a consistency checker, a problem solver, and a results analyzer. The instance interpreter is for interpreting the data collected from the Template Creators (or Editors) and formatting it into DSP Template instances according to the DSP ontologies. The generated template instances and module instances are stored in the Knowledge Base. The search engine is connected to the Knowledge Base to provide ontological semantic-based knowledge retrieval. Consistency checking is facilitated through a consistency checker together with the JESS Reasoner—the Rule Engine for the JavaTM Platform provides rule-based intelligence inference. The problem solver is connected to DSIDES for solving the DSPs. DSIDES (Decision Support in the Design of Engineering Systems) is a tailored computational environment that supports the execution of the decision support problem templates. DSIDES is invoked when a template executer executes a DSP template. The result analyzer is included to help users, especially Template Implementers analyze the results produced by the problem solver, DSIDES. MATLAB and its features are used to carry out data visualization through ternary plots and scatter plots. These plots are used for visualizing the DSP results and further carrying out solution space exploration. Therefore, MATLAB and its features are linked to CB-PDSIDES for solution space visualization, solution space exploration, and post-processing. The virtual and physical layers in CB-PDSIDES support data storage and retrieval for multiple users connected via the cloud, thereby establishing a networked collaborative environment. As noted by Wu and co-authors [2], through virtualization, the computing resources of CB-PDSIDES like the DSP templates, solvers, post-processing resources, data storage, etc., are separated from physical computing hardware and reallocated dynamically to the different applications based on the needs of users. Through this unified computing architecture of CB-PDSIDES, we are able to support multiple tenants through a single instance of the platform PDSIDES; defined as multi-tenancy [2]. These features of the computing architecture of cloud-based PDSIDES differentiate it from the other web-based services. Since CB-PDSIDES is a highly flexible SOA-based platform, there are some challenges for the realization of it. In the following sub-sections, we discuss these challenges and propose some open questions worthy of future exploration.

7.2.2 *Service Modeling*

The first big challenge is to "servitize" the declarative/procedural/integrated knowledge in CB-PDSIDES, so that everything in the cloud is a piece of service that can satisfy a particular kind of need. Decision-related knowledge (i.e., declarative/procedural/integrated knowledge) has different structures and it is difficult to describe them in a unified standard way to facilitate management. We note that the Web Services Description Language (WSDL) [4], which is an interface definition language for describing the functionality of a web service, may be an alternative language to model the services in CB-PDSIDES. WSDL consists of two parts, an abstract part that describes the operational behavior of the web service, and a concrete part that describes how and where to access a service implementation. The limitation of WSDL is anchored in that and it only tells what messages go in and come out from a service but does not provide the terms for describing the semantics of the service which results in the difficulty of achieving transparency and trustworthiness (especially when the service is used to support decisions). Other challenges related to service modeling include (but are not limited to) the classification and granularity partitioning of the services. There is a need for CB-PDSIDES to support the flexible classification of services to facilitate searching. A service may belong to multiple

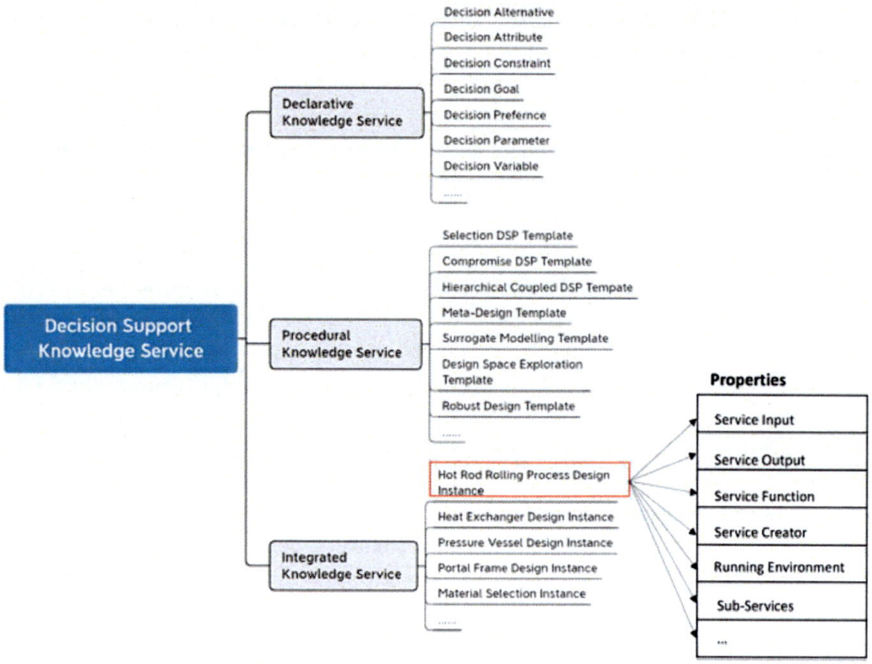

Fig. 7.4 Classification of the services of CB-PDSIDES based on knowledge types

categories. For example, a constraint that is considered in the design of a pressure vessel can be classified as declarative knowledge from a knowledge type perspective, and can also be classified into mechanical products from a product perspective. The classification of the services based on knowledge types is shown in Fig. 7.4, which captures certain service properties from a knowledge type perspective, e.g., service input, service output, running environment, sub-services, etc. In addition to this, some domain-specific classification schemes need to be implemented in CB-PDSIDES, so that more domain properties can be captured and more service indexes can be established to enable flexible searching. Granularity partition means that the level to which a service is partitioned into components. Granularity partition has impacts on the customizability of a service (which will be discussed in Sect. 7.2.2).

Key questions worthy of further investigation in this thrust are summarized as follows:

- What is a description scheme for modeling the services in CB-PDSIDES so that semantic richness, transparency, and trustworthiness for decision support can be achieved?
- How to provide flexible, multidimensional classification and visualization of services in CB-PDSIDES to facilitate searching and navigation of the services?
- How can service granularity partition be controlled to enable flexible customization?

7.2.3 Service Customization

The second big challenge we foresee in the realization of CB-PDSIDES is to enable cloud-based mass service customization, so that designers can configure, select, and utilize the decision support services according to their preferences. Providing customized services is important for CB-PDSIDES to attract users with individualized needs and help them succeed in dynamic and competitive market environments. The difficulty in enabling mass customization of decision support service is anchored in the lack of customized service design methods for designing and evaluating the services offered to customers. One possibility is to borrow the idea of product customization and extend product family design methods to service family design. A product family is a set of products that share a common product platform and differentiate from each other by some unique modules or components [5]. Similarly, a service family can be defined as a set of services which share a service platform and differentiate each other with unique modules or components [6]. A service module consists of a set of components certain services. In CB-PDSIDES where the services are related to decision support, in order to design a service family, it is necessary to identify the common modules that form the service platform, and unique modules that certain services must possess, and variant modules that are optional for specific services. A good design of a service family not only provides the flexibility for mass customization, but also ensures cost-effectiveness (in terms of computing resource consumption and maintenance). Essentially, this means making a trade-off in the selection of a platform level, to determine the number of common and variant modules, as shown in Fig. 7.5. If the platform level is high (i.e., the commonality among services is high), then the cost of modules, as well as the interfaces for connecting them, is low, while customers' preference loss is high because

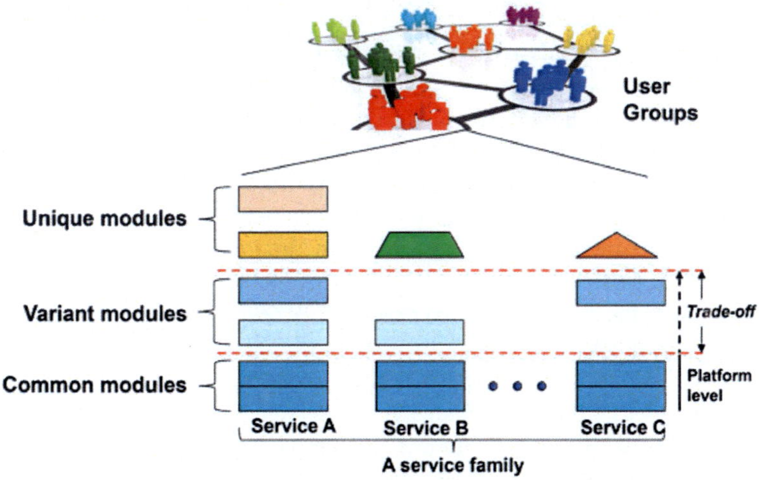

Fig. 7.5 Trade-off in service platform level selection (revised from [6])

of the lack of uniqueness in the service. On the hand low platform levels, with low commonality among services, will increase the module and interface costs while decreasing the customers' preference loss.

Key questions worthy of further investigation in this area are summarized as follows:

- How to model service families of design processes, decision workflows, and individual decisions so that service customers can customize them based on personalized needs?
- How to manage the trade-off between commonality and flexibility when designing service families in CB-PDSIDES?
- What is a appropriate pricing method for the customized services considering the problem that needs to be solved and the computational resources that need to consume?

7.2.4 Intelligent Service Composition

The third big challenge is to enable the intelligent composition of service packages from module level to individual decision level, and from individual decision level to decision workflow level, so that the service components can be scaled up to fulfill customer requirements. Wu and coauthors [7] define service composition as a process of composing services with different functions into a larger service to meet user request when no single service can fulfill those requests. They further give an example of service composition as: if the output parameters of Service A can be used as the input parameters of Service B, then A and B can be connected as a new service with input parameters that are the same as the input parameters of A and output parameters that are the same as the output parameters of B. In the context of CB-PDSIDES, service composition is a necessary functionality because solving of complex system design problems typically requires complex decision workflows which consist of a set of decisions which require a set of modules, and no single service related to a module or decision can solve the complex problem independently. Typically, a service composition process takes several steps including partitioning of the requirement into sub-requirements, identification of services that are related to the sub-requirements, connecting the related services as a chain or network which then forms a service package, evaluation of the service package, and improving it through service reconfiguration. We expect that there will be a huge number of services uploaded to the service pool in the cloud by service providers and these services need to be formed as packages and ready to deliver to service customers rapidly so as to improve user experience. To realize this, there is a need for an intelligent service composition function in CB-PDSIDES which deeply "understands" the requirements of service customers and intelligently composes service packages for satisfying the requirements, and the quality of the service packages generated by intelligent composition is as good as or better than the ones generated by manual composition. In Fig. 7.6, we show a conceptual process for intelligent service composition. Given the customer

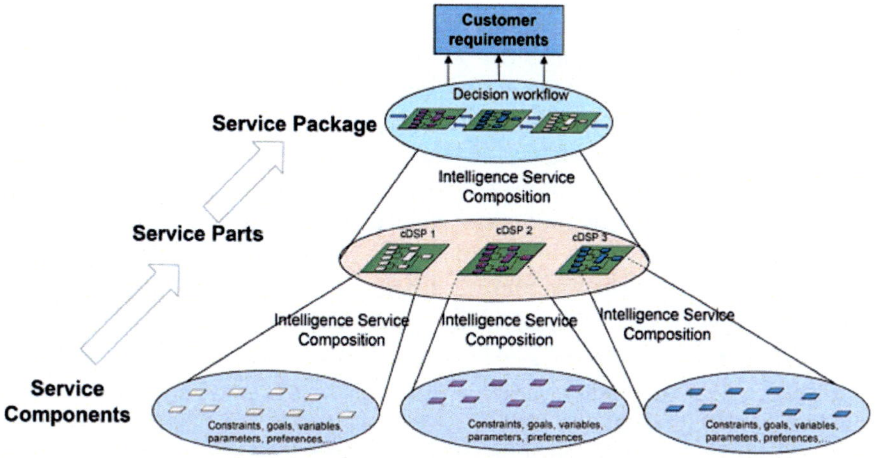

Fig. 7.6 Bottom-up intelligent composition of services in CB-PDSIDES

requirements, CB-PDSIDES "understands" that to meet the requirements on an alternative(s) to formulate a decision workflow that consists of three coupled cDSPs, and then intelligently compose a service package involving these three cDSPs and deliver it to the customer. As shown in Fig. 7.6, the intelligent service composition process is a two-level, bottom-up process. It initiates by searching all the related cDSP components including constraints, goals, variables, parameters, etc., from the service pool, then assembles the components using cDSP templates, and finally, the assembled cDSPs templates are linked as a decision workflow.

Key questions worthy of further investigation in this area are summarized as follows:

- How can machine learning techniques be applied to the intelligent composition of decision support services that involve the composition of decision workflows from cDSP/sDSP templates, and the composition of cDSP/sDSP templates from template modules?
- How can compatibility be defined in service composition that facilitates consistency checking after a service package is composed?

7.2.5 *Smart Service Provider-Seeker Matching*

The fourth big challenge that we see in CB-PDSIDES is to enable smart service provider-seeker matching so that both service seekers' and providers' satisfaction is maximized. A service provider who uploads some declarative, procedural, or integrated knowledge to CB-PDSIDES hopes that their knowledge can be "best" used to support decisions and create value. And a service seeker who searches services in CB-PDSIDES hopes to find the "best" knowledge service for supporting their design decisions. It is important to build a provider-seeker matching mechanism in CB-PDSIDES so that both sides are happy and the ecosystem can sustain and grow. Mandhan and co-authors [8] propose to use the Gale-Shapley algorithm for matching designers and 3D printing services providers. In their problem, both sides (i.e., designers and manufacturers) consist of decision-makers who can determine to whom the printing service is provided and from whom the printing service is selected, and both try to maximize their utility when making decisions. A similar situation happens in CB-PDSIDES as well. Both the knowledge service providers and seekers are decision-makers who can make decisions on the services. The Gale–Shapley algorithm is an alternative for building the matching mechanism in CB-PDSIDES, as shown in Fig. 7.7. Service providers can be those template creators who create decision templates and want to "sell" them in CB-PDSIDES. Service seekers can be those who have problems and want to "buy" the decision templates in CB-PDSIDES and implement them. The core difficulty is to set up the utility functions that measure the "happiness" of service providers and seekers, respectively, and use them to order the preferences over the candidate options. Since optimal decisions are sometimes difficult to achieve in the matching, the pursue of satisficing or compromise decisions may be appropriate for determining the matching mechanism.

Key questions worthy of further investigation in this area are summarized as follows:

- How to quantify the "happiness" or utility of service providers and service seekers when they make decisions about matching each other in CB-PDSIDES?
- What is a provider-seeker matching mechanism that maximizes the "happiness" or utility of both the service providers and seekers simultaneously?
- How can a provider-seeker matching problem be formulated using the Compromise Decision Support Problem construct so that satisficing solutions can be identified which are acceptable to both the service providers and seekers?

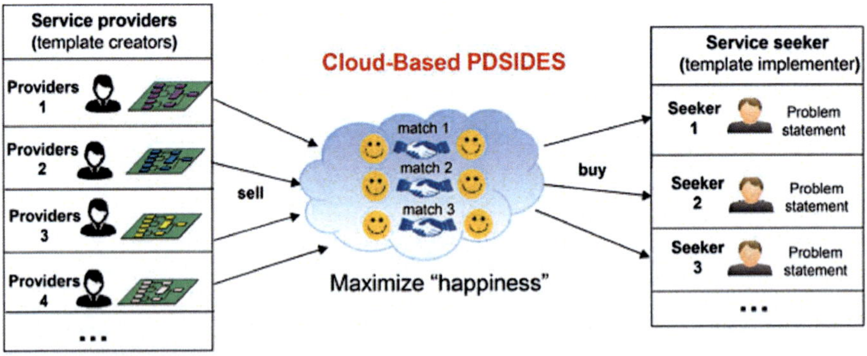

Fig. 7.7 Smart service provider-seeker matching in CB-PDSIDES

7.2.6 Mechanism for Design Collaboration (Co-Design)

The fifth major challenge that we envision is the capability to carry out "co-design" in a secure fashion. Design engineering in the era of Industry 4.0 necessitates co-design, which we define as the capability of a network of participants in the value chain, including material scientists, systems designers, software developers, and end customers, to come together and share material/product/manufacturing process/market data, information, knowledge, and resources instantly and in an integrated fashion, thereby to collaborate and facilitate a cost-effective co-creation of value supporting open innovation. Using CB-PDSIDES, collaborating users from different parts of the world should be able to use the cloud securely and systematically and: (a) develop and integrate digital models, (b) formulate co-design problems, (c) explore co-design solutions that meet end requirements, and (d) manage uncertainty and explore robust solutions for the co-design problems. Knowledge-based co-design guidance and decision support should also be available to the users. Integrated knowledge, i.e., integration of declarative and procedural knowledge is captured via design methods and frameworks in CB-PDSIDES. Design methods such as goal-oriented inverse design and inductive design exploration methods incorporated in CB-PDSIDES facilitate the co-design of materials, products, and manufacturing processes meeting goals and requirements at multiple levels for different stakeholders. Together with these methods, Secure Co-Design (SCD) frameworks that use information theory and game theory protocols to identify co-design solutions while preserving the confidentiality of the information shared between the participating design collaborators is a key requirement in CB-PDSIDES. SCD frameworks enable secure collaborative designs among different stakeholders or designers so that the system is jointly designed without leaking the privacy information of different stakeholders. From the context of integrated materials and product design, this can be viewed as a situation where product components are designed by one design group and the corresponding materials are designed by material scientists who work as the second group. The situation demands information and knowledge sharing that needs to be managed. The issue of privacy and secure co-design collaborations are critical in this situation. CB-PDISIDES can offer cloud-based co-design through SCD frameworks in which components are designed by one organization and materials are designed by another organization. As an example, we can see this using the hot rolling example addressed in Chap. 4. The cDSP for the rod (product) is formulated by product designers who have specific mechanical property requirements for the product. The predictions of these mechanical properties will be using specific product-level simulation models that are proprietary. The microstructure requirements for the product at different stages are formulated using the cDSP construct by material scientists. They will also have proprietary models for microstructure evolution which they may not be interested in sharing with the product design team. In a similar sense, the product designers do not wish to share all the information in the product-level cDSP with material scientists. However, using the cloud-based co-design functionalities, both parties would like to jointly design the product and

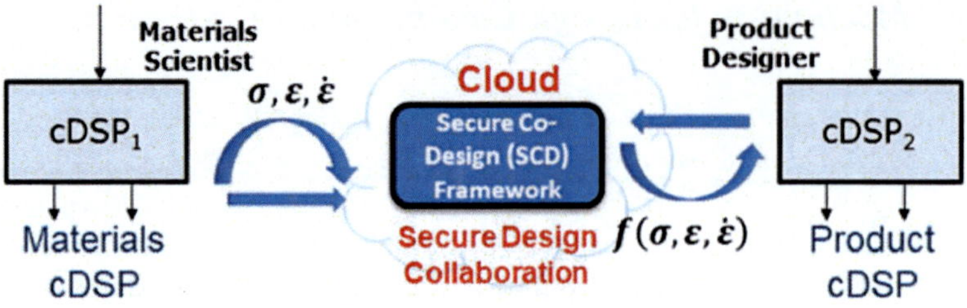

Fig. 7.8 Illustration of secure co-design using CB-PDSIDES

the material. The illustration of secure co-design using CB-PDSIDES is shown in Fig. 7.8. This requires addressing additional challenges, such as:

- How to collaboratively author the workflow templates and check for consistency among the collaborating stakeholders/agents
- How to model the individual decision-making strategies, their interactions, and drive the collective system interaction to meet the overall requirements?
- How to facilitate privacy-preserving collaboration in integrated products and materials design?

7.3 Broader Applications

CB-PDSIDES is developed as a platform for cloud-based design decision support for multiple applications and problem requirements. CB-PDSIDES can be used to formulate complex systems design problems by modeling the workflow of compromise and selection decisions from a Decision-Based Design perspective. Further solutions that satisfice multiple requirements can be explored. In Fig. 7.9, we show broader applications where CB-PDSIDES is used. The applications are addressed from a Cyber-Physical standpoint with the human in the loopHuman-Cyber-Physical systems. We discuss each of these applications briefly in this section.

7.3.1 Applications to Cyber-Biophysical Systems

In the Cyber-Biophysical Systems domain, we explore the application of CB-PDSIDES to integrate sensing, computational, and communications networks with biology. Example problems such as head and neck injury-based design exploration for vehicular crashworthiness [9, 10], design of a ventricular shunt for hydrocephalus [11] are used to illustrate the efficacy of the platform for the design of cyber-biophysical systems. In these example problems, we use the functionalities of CB-PDSIDES to carry out cloud-based problem formulation and simulations (car crash

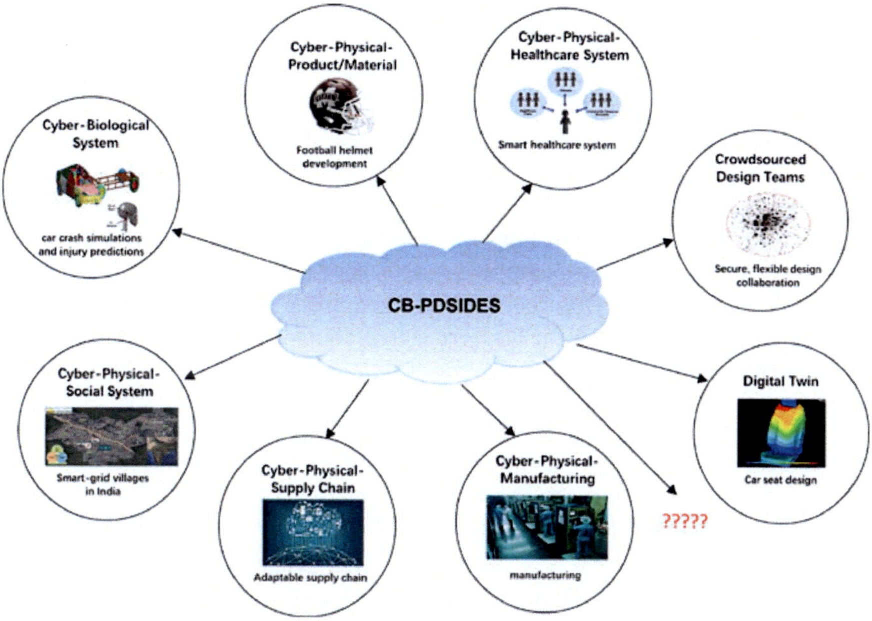

Fig. 7.9 Broader applications of CB-PDSIDES

simulations and injury predictions for crashworthiness example, CSF flow rate simulations for the shunt design example), the development of surrogate models capturing the responses of interest (head injury criteria surrogate models and neck injury criteria surrogate models for the vehicular crashworthiness example, shunt flow rate surrogate model (for the shunt design example, etc.), the formulation of robust design problems, solution space exploration, and uncertainty management; see Fig. 7.10 [12, 13].

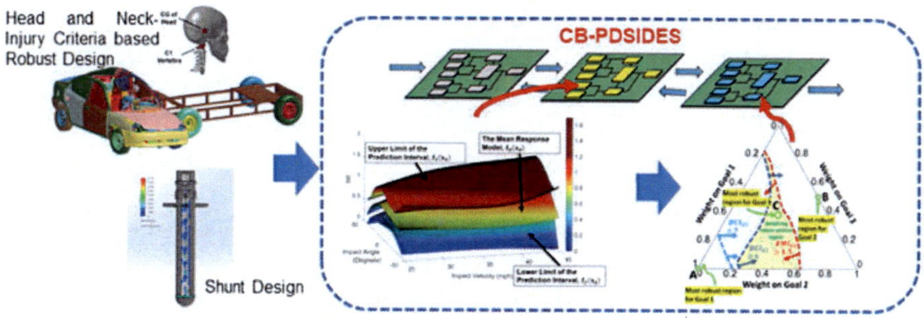

Fig. 7.10 CB-PDSIDES for Cyber-Biophysical Systems

7.3.2 Applications to Cyber-Physical-Product/Material Systems

From the context of the Integrated Computational Materials Engineering (ICME) domain, we are interested in exploring the application of CB-PDSIDES to address the integrated design of materials and products. This involves designing the material to meet system-level requirements. An example is the design of American football helmets. The CB-PDSIDES platform and associated design frameworks such as CEF and GoID are used to design components (composite shell and foam liner) of an American football helmet [14]. The goal is to design the products, sub-components, and materials using multi-scale modeling efforts and systems-based robust co-design techniques in an integrated fashion.

7.3.3 Applications to Cyber-Physical-Manufacturing Systems

In the context of Industry 4.0, we are interested in exploring the design of manufacturing systems by addressing the distributed and networked nature of manufacturing processes and associated products. CB-PDSIDES will be used to facilitate seamless data, information, and knowledge sharing among the different physical and cyber components of the manufacturing system and the stakeholders. A computational framework for designing the dynamic management of such networked manufacturing systems is proposed [15]. The integration of the framework and constructs to CB-PDSIDES is envisioned to design cyber-physical-manufacturing systems that are adaptable.

7.3.4 Applications to Cyber-Physical-Social Systems

CB-PDSIDES for automated micro-enterprise design and analysis to achieve sustainable rural development is addressed as an application of Cyber-Physical-Social systems, described by Yadav and co-authors [16]. Yadav and co-authors [16] present a computational framework incorporating three constructs that support this application, namely, a Village Level Baseline Sustainability Index, a Dilemma Triangle, and Village Level System Dynamics. Using these constructs, a dialog is facilitated among the stakeholders, namely, corporate social responsibility (CSR) investors and social entrepreneurs. These stakeholders are geographically dispersed and an increase in the number of stakeholders demands a need for effective communication and collaboration in order to come up with sustainable solutions (micro-enterprises) worthy of further investment. Using CB-PDSIDES, the stakeholders will be able to direct attention to issues and challenges that are typically ignored or missed while solving complex social problems.

7.4 CB-PDSIDES for Design Engineering 4.0

Systems realization in the age of Industry 4.0 demands a new design paradigm that embodies the distributed and networked aspects of systems. This paradigm, defined as Design Engineering (DE) 4.0, requires a "human-cyber-physical view of the systems realization ecosystem" to accommodate the drivers of Industry 4.0 (IoX) and provide an open ecosystem for the realization of complex systems. CB-PDSIDES for DE 4.0 address future research from the perspectives of human, business, system, and cybernetics and are driven by IoX, see Fig. 7.11. Nellippallil and co-authors [3] identify the core competencies needed for CB-PDSIDES users for realizing DE 4.0. These competencies are:

- The capability to integrate models and simulation tools spanning different processes and length scales (typically defined as vertical and horizontal integration from the context of integrated design of materials, products, and manufacturing processes),
- The capability to define computational workflows involving decision-making, spanning multiple activities and users,
- The capability to define modular, reusable sub-workflows for specific processes,
- The capability to connect to external databases about materials, products, and processes,
- The capability to provide knowledge-guided assistance for different types of users in design-related decision-making,
- The capability to carry out collaborative, multidisciplinary design, and privacy control,
- The capability to manage complexity (reduce the cost of computation via surrogate models/meta models),
- The capability to explore and visualize the design and solution space,
- The capability to carry out dynamic and cost-efficient reconfiguration and integration of design decision templates to explore different robust design strategies (meta-design to deliver robust products).

The key areas for future research using CB-PDSIDES include [17]: (a) ***Design for*** User Experience, (b) ***Design of*** Human-Cyber-Physical Systems, (c) ***Design with*** Smart Sensing and Artificial Intelligence Technologies, and ***Design as*** Strategic Engineering. From the human perspective, CB-PDSIDES will be used to support the design for user experience not just for end customers in the value chain, but also for multiple stakeholders involved across the product realization cycle. This results in the Internet of People for DE 4.0. The primary challenge is to address the changing user preferences and how different platform users interact with the services, products, and between themselves using CB-PDSIDES. From the business perspective, CB-PDSIDES will be used to model the interactions and decisions between businesses (B2B), between business and consumers (B2C), and between consumers (C2C). This results in the Internet of Commerce for DE 4.0. From the cyber perspective,

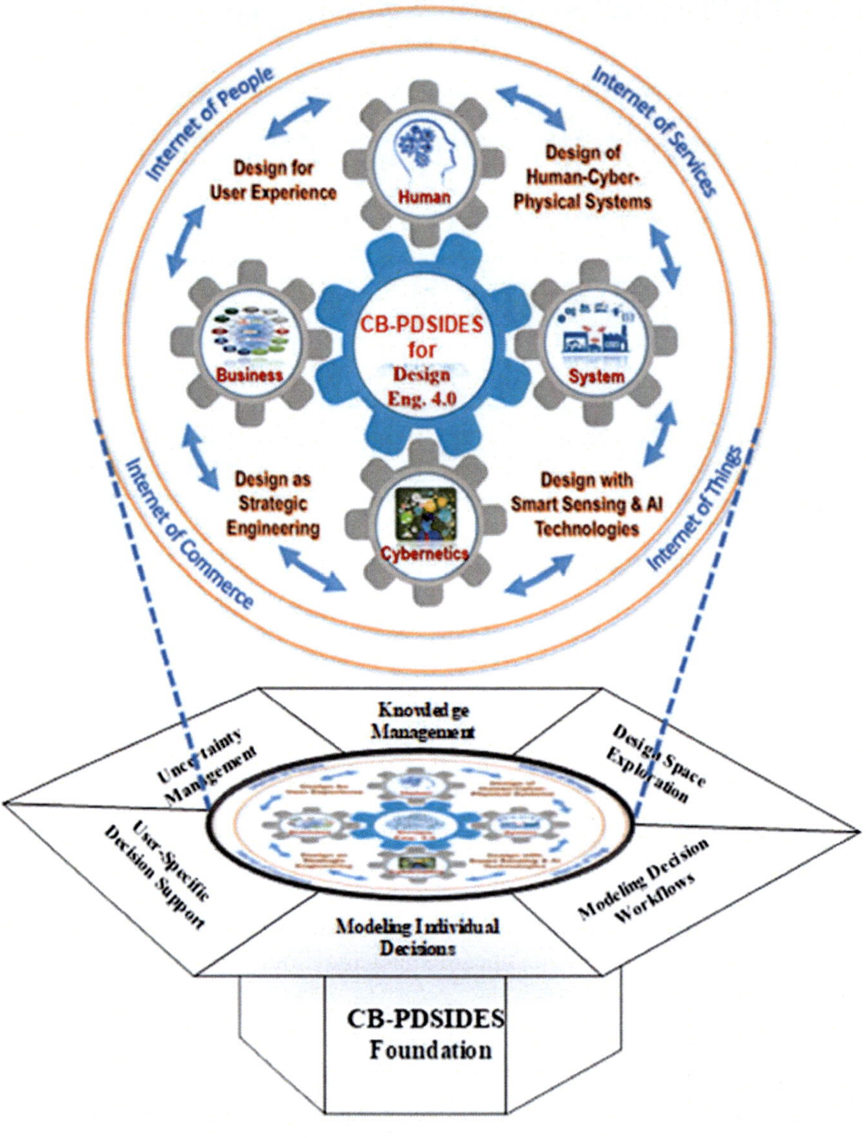

Fig. 7.11 CB-PDSIDES foundations for design engineering 4.0

the focus will be on incorporating the advancements in machine learning, artificial intelligence, and data-driven decision-making approaches into CB-PDSIDES to facilitate informed decision-making under uncertainty. This results in the Internet of Things for DE 4.0. From the systems perspective, the focus using CB-PDSIDES will be to integrate the physical domain with the cyber domain keeping humans in the loop, thereby implementing self-evolving human-cyber-physical systems of the future. This results in the Internet of Services for DE 4.0. We hypothesize that CB-PDSIDES for DE 4.0 needs to address these drivers of Internet of People, Commerce, Things, and Services and their respective associated challenges. The foundations of CB-PDSIDES offer core functionalities that need to be leveraged and expanded to address the DE 4.0 drives. The core functionalities include:

- ***Knowledge Management using CB-PDISES for DE 4.0.*** Decision-related knowledge (including procedural, declarative, and integrated knowledge) associated with the design of cyber-physical-social systems can be captured using decision support problem templates. The decision support ontologies will allow the users to search, share, and reuse the knowledge as systems evolve and requirements change. More details on the decision support ontologies are available in Chaps. 3 and 4 of this monograph.
- ***Individual Decision Formulation using CB-PDSIDES for DE 4.0.*** Using CB-PDSIDES and associated Decision Support Problem (DSP) constructs individual decisions related to cyber, physical, social domains can be formulated by collaborating designers from different parts of the world. Co-design, that is, the ability of designers/users to come together and share their information, knowledge, and resources instantly and in an integrated manner, to collaborate to create new products, processes, and services is facilitated. Using CB-PDSIDES designers can provide instant feedback, and this will allow the individual decisions to be flexible and adaptable to changes. Selection decisions that involve a choice among a number of possibilities considering a number of measures of merit of attributes, and compromise decisions that involve determining the "right" values of design variables to describe the best satisficing solution with respect to constraints and multiple goals, can be carried out via the cloud using the selection DSP (sDSP) and the compromise DSP (cDSP) constructs, respectively. More details on individual decision formulation using DSP are available in Chap. 4 of this monograph.
- ***Decision Workflows Formulation using CB-PDSIDES for DE 4.0.*** Using CB-PDSIDES and the PEI-X diagram, design decision workflows that are critical in meta-design of design processes can be formulated. These decision workflows can be used to represent design processes for complex cyber-physical-social systems. They can be further analyzed by carrying out an exploration of design alternatives before actual implementation. More details on decision workflows are available in Chaps. 4 and 5 of this monograph.
- ***Solution Space Exploration using CB-PDSIDES for DE 4.0.*** Using CB-PDSIDES, designers are able to explore the solution space by following a formal procedure that centers on the cDSP construct. The procedure takes design requirements as input and gives satisficing design specification as output. Steps include clarification of the design event, definition of the problem, identification of theoretical and empirical models that are available, development of surrogate models, formulation of the cDSP, executing the cDSP and carrying out solution space exploration, visualization, and trade-offs. The solution space exploration functionality ensures that the output design specification is acceptable to the designer. All these functionalities are critical for realizing systems and addressing the drivers

of DE 4.0. More details on solution space exploration are available in Chaps. 4, , 5, and 6 of this monograph.

- ***Uncertainty Management using CB-PDSIDES for DE 4.0***. Dynamically managing uncertainty and complexity across process chains is one of the critical functionalities needed to address the business and cyber aspects of DE 4.0. In CB-PDSIDES, designers are able to explore robust solutions for problems by managing different sources of uncertainties. This is carried out using the robust design metrics, Design Capability Indices (DCI), and Error Margin Indices (EMI), respectively [18, 19]. Using these functionalities, designers are able to identify satisficing solutions that are relatively insensitive to uncertainty. In CB-PDSIDES, designers have access to robust concept exploration methods, such as the goal-oriented inverse design (GoID) method, and robust concept exploration method (RCEM) to carry out robust concept exploration of process chains by managing uncertainty. More details on uncertainty management are available in Chap. 6 of this monograph.
- ***User/Activity Specific Decision Support using CB-PDSIDES***. To facilitate co-design and achieve the vision of DE 4.0, there is a need to tailor decision support by accommodating different types of users in a variety of activities. CB-PDSIDES supports three types of users for customized decision support, namely, *domain experts* or *template creators* who are responsible for creating decision templates in original design activities, *senior designers* or *template editors* who are responsible for editing (or tailoring) existing decision templates in adaptive design activities, and *novice designers* or *template implementors* who are responsible for executing existing decision templates in variant design activities. More details on user/activity specific decision support are available in Chap. 4 of the monograph.

7.5 Closing Comments

The paradigm shift in the age of Industry 4.0 requires a "human-cyber-physical view of the systems realization ecosystem", and platformization for augmenting the role of designers as decision-makers are believed to be critical for manufacturing enterprises to succeed in this revolution. In this monograph, we describe how a knowledge-based platform for decision support in the design of engineered systems is architected to respond to this paradigm shift. In Chap. 1, the definitions of several background concepts (including Industry 4.0, Design Engineering 4.0, and the Industrial Brain) are reviewed, and the motivation, requirements, and basic architecture of the PDSIDES platform are discussed. Even though big data is one of the key features of Industry 4.0 and Design Engineering 4.0, we believe what really supports designers making design decisions is knowledge and there is a need for a platform to make full use of knowledge to provide decision support. Such a platform should have functionalities including knowledge management and reuse, formulation of decisions and decision workflows, solution space exploration, uncertainty management, and user/activity specific decision support, in order to meet dynamic needs. In Chap. 2, we review and discuss the constructs for enabling those functionalities of PDSIDES. The constructs include the DSPs for formulating decisions, the PEI-X for constructing decision workflows, the procedure and indicators for managing uncertainty and making robust decisions, and ontologies for knowledge management and reuse. In Chap. 3, we discuss the development of the key ontologies for building the knowledge base for PDSIDES. First, an ontology is developed to represent the knowledge related to selection decisions, and its utility is tested using a light switch cover plate material selection example. Second, an ontology is developed to represent the knowledge related to compromise decisions and its utility is tested using a pressure vessel geometry design example. Third, an ontology is developed to represent the knowledge related to hierarchical coupled decisions (including selection and/or compromise decisions) and its utility is tested using a portal frame design example. These three ontologies form the core structure of a knowledge base for PDSIDES to support decisions in design. In Chap. 4, a prototype of PDSIDES is developed, and three types of users and their specific activities are defined. Template creators are expert designers who create decision templates in original design, template editors are senior designers who tailor existing decision templates in adaptive design, and template implementers are novice designers who specify the parameters in decision templates and execute them in variant design. The usefulness of the prototype of PDSIDES is tested using a hot rod rolling system design example. In Chap. 5, the functionality of knowledge-based meta-design of decision workflows is added to PDSIDES and the associated ontology is developed, and a heat exchanger design example is used to test this functionality. In Chap. 6, the functionality of knowledge-based robust design space exploration is added to PDSIDES and the associated ontology is developed, and it is tested using the hot rod rolling system design example. We believe the knowledge-based decision support functionalities defined in this monograph will benefit enterprises who have interests in the design of complex systems in the context of Industry 4.0.

Putting the knowledge-based functionalities into the cloud is the future direction for PDSIDES, that is to create the so-called "CB-PDSIDES" for delivering cloud-based decision services. The key to CB-PDSIDES is the SOA architecture. We identify several challenges for realizing such a SOA architecture in the context of design decision support, including service modeling, service customization, intelligent service composition, and smart service provider-seeker matching. We also propose the related research questions for tackling each of these challenges. The answers to these questions, we believe, will shape the future of decision support with a "human-cyber-physical view of the systems realization ecosystem."

As technology advances, we anticipate that autonomy will be shared between cyber systems and humans in Human-Cyber-Physical systems. However, as the decision-maker, a human still needs to decide when a "Human supports a Computer" or when a "Computer supports a Human" as the system evolves over time. It is the purview of the designer to select the transfer of autonomy from cyber-operators to human operators and vice versa. In that sense, how to provide proper support to designers to make such meta-level decisions is a new topic, that merits another monograph. Stay tuned!

References

1. Thames, L., & Schaefer, D. (2017). *Cybersecurity for industry 4.0.* Springer.
2. Wu, D., Rosen, D. W., Wang, L., & Schaefer, D. (2015). Cloud-based design and manufacturing: A new paradigm in digital manufacturing and design innovation. *Computer-Aided Design, 59*, 1–14.
3. Nellippallil, A. B., Ming, Z., Allen, J. K., & Mistree, F. (2019). Cloud-based materials and product realization—fostering ICME via industry 4.0. *Integrating Materials and Manufacturing Innovation, 8*(2), 107–121.
4. Christensen, E., Curbera, F., Meredith, G., & Weerawarana, S. (2001). *Web services description language (WSDL)* W3C Note 15.
5. Halman, J. I. M., Hofer, A. P., & van Vuuren, W. (2006). Product platform and product family design, In T. W. Simpson, Z. Siddique, J. R. Jiao (Eds.), *Product Platform and Product Family Design.* Chapter 3, Springer.
6. Moon, S. K., Simpson, T. W., Shu, J., & Kumara, S. (2008). A method for platform identification to support service family design. *International Journal of Services Operations & Informatics, 3*(3/4), 294–317.
7. Wu, Z. (2014). *Service computing: Concept.* Academic Press.
8. Mandhan, N., Thekinen, J., Lo, A., & Panchal, J. H. (2016). Matching designers and 3D printing service providers using gale-shapley algorithm. In *Proceedings of the eleventh international symposium on tools and methods of competitive engineering.*
9. Nellippallil, A. B., Berthelson, P. R., Peterson, L., & Prabhu, R. K. (2022). A computational framework for human-centric vehicular crashworthiness design and decision-making under uncertainty. *ASCE-ASME Journal of Risk and Uncertainty in Engineering Systems Part B: Mechanical Engineering.* https://doi.org/10.1115/1.4053515.
10. Nellippallil, A. B., Berthelson, P. R., Peterson, L., & Prabhu, R. K. (2022). Robust concept exploration of driver's side vehicular impacts for human-centric crashworthiness. In R. Prabhu, M. Horstemeyer (Eds.), *Multiscale Biomechanical Modeling of the Brain* (pp. 153–176), Academic Press. ISBN 9780128181447, https://doi.org/10.1016/B978-0-12-818144-7.00002-5.
11. Sharma, G., Nellippallil, A. B., Yingling, R., Lee, N. Y., Shores, A., Miralami, R., Stone, T. W., & Prabhu, R. K. (2021). Multi-objective robust design exploration of a canine ventricular shunt for managing hydrocephalus. *47th Design Automation Conference (DAC).* Virtual, Paper Number DETC2021-67353.
12. Nellippallil, A. B., Berthelson, P. R., Peterson, L., & Prabhu, R. (2020). Head and neck injury based robust design for vehicular crashworthiness. *ASME design automation conference*, Paper Number IDETC2020-22539.
13. Yingling, R., Nellippallil, A. B., Register, M., Hannan, T., Simmons, J., Shores, A., & Prabhu, R. K. (2020). Multi-objective design exploration of a canine ventriculoperitoneal shunt for hydrocephalus. In *Proceedings of the international design engineering technical conferences and computers and information in engineering conference*, Paper Number DETC2020-67353.
14. Fonville, T. F., Nellippallil, A. B., Horstemeyer, M. F., Allen, J. K., & Mistree, F. (2019). A goal-oriented, inverse decision-based design method for multi-component product design. *ASME Design Automation Conference,* Paper Number DETC2019-97388.
15. Milisavljevic-Syed, J., Allen, J. K., Commuri, S., & Mistree, F. (2019). Design of networked manufacturing systems for indusrty 4.0. In *CIRP manufacturing systems conference 2019.*
16. Yadav, A., Das, A. K., Allen, J. K., & Mistree, F. (2019). A computational framework to support social entreprnuers in creating value for rural communities in India. *ASME design automation conference,* Paper Number DETC2019-97375.
17. Jiao, R., Commuri, S., Panchal, J., Milisavljevic-Syed, J., Allen, J. K., Mistree, F., & Schaefer, D. (2021). Design engineering in the age of industry 4.0. *Journal of Mechanical Design, 143*(7), 070801.

18. Choi, H. J., Austin, R., Allen, J. K., Mcdowell, D. L., Mistree, F., & Benson, D. J. (2005). An approach for robust design of reactive power metal mixtures based on non-deterministic micro-scale shock simulation. *Journal of Computer-Aided Materials Design, 12*(1), 57–85.
19. Nellippallil, A. B., Mohan, P., Allen, J. K., & Mistree, F. (2020). An inverse, decision-based design method for robust concept exploration. *Journal of Mechanical Design 142*(8), 081703.

Index

© The Editor(s) (if applicable) and The Author(s), under exclusive license to Springer Nature Switzerland AG 2022
Z. Ming et al., *Architecting A Knowledge-Based Platform for Design Engineering 4.0*,
https://doi.org/10.1007/978-3-030-90521-7